T0172506

40 Top Tools for Manufacturers

40 Top Tools for Manufacturers

A Guide for Implementing Powerful Improvement Activities

Walter J. Michalski, Ed.D.

edited by
Dana G. King, M.A.

Productivity Press

Copyright ©1998 by Productivity Press

All rights reserved. No part of this book may be reproduced or utilized in any form or by any means, electronic or mechanical, including photocopying, recording, or by any information storage and retrieval system, without permission in writing from the publisher.

Cover illustration by Ana Capitaine

Library of Congress Cataloging-in-Publication Data

Michalski, Walter J.
    40 top tools for manufacturers : a guide for implementing powerful improvement activities / Walter J. Michalski ; edited by Dana G. King.
        p.   cm.
    A companion volume to the author's Tool navigator (1997).
Includes index.
    ISBN 1-56327-197-4
    1. Total quality management.   2. Group problem solving.
3. Manufacturers—Computer programs.   4. Manufacturers—Charts, diagrams, etc.   I. King, Dana G.   II. Michalski, Walter J. Tool navigator.   III. Title.
HD62.15.M528   1998
658.5′62—dc21   97-48352
                                                              CIP

10 9 8 7 6 5 4 3 2

Contents

Publisher's Message

As a companion to our recently published Master Guide, *40 Top Tools for Manufacturers—A Guide for Implementing Powerful Improvement Activities* provides a carefully selected package of tools for a specific audience, the manufacturer. The 40 tools are arranged in four stages: manufacturing process, cycle time, variability, and problem solving, and they appear in the same easy-to-read format found in the *Tool Navigator™—The Master Guide for Teams*. These manufacturing tools can be used to assist teams in the proper selection, sequencing, and application of major total quality methods (TQM) such as just-in-time (JIT), total productive maintenance (TPM), process mapping (P-M), cycle time management (CTM), and many other methods. This is the first book in a series of problem-solving and quality/process improvement tool books Productivity Press will be developing based on the *Tool Navigator™*. These streamlined books will use tools that focus on specific needs and areas of interest for problem-solving teams. For example, there will be tool books tailored for engineering, marketing, and research. Or, they will be aimed toward different problem-solving phases or organized by tool classifications, such as idea generating, team building, and data collecting.

Acknowledgments

I would like to give my special thanks to my wife, Giovanna, who gave valuable advice during the drafting of many of these tools. Many sincere thanks to my daughter, Dana Giovanna King, for spending many long hours editing and revising my drafts, and appreciation for the contributions of my sons—James Walter for his special counsel on all administrative matters and Anthony Peter for his typing assistance. I am also indebted to my brother, Peter Michalski, for sharing his computer expertise, insights, and constructive evaluations.

I would like to also thank the originators or sources of many of the tools presented in this book. I have made every effort to identify and credit them and offer my sincere apology if I have overlooked anyone.

Finally, I want to thank the team at Productivity Press, headed by Diane Asay, Editor in Chief, and the development editor, Gary A. Peurasaari, who provided me with suggestions and technical support.

Walter J. Michalski, Ed.D.

Introduction

40 Top Tools for Manufacturers—A Guide for Implementing Powerful Improvement Activities is the response to many requests I have had from TQM practitioners for a listing of the appropriate manufacturing tools for problem-solving teams. The tools are taken from the *Tool Navigator™—The Master Guide for Teams* handbook, also published by Productivity Press, Inc., that contains 222 tools. *The Tool Navigator* provides teams with an easy-to-read format of powerful tools for problem-solving and quality/process improvement activities. I wrote it as a result of my constant search for additional, special, or more appropriate tools to use in order to enhance my team facilitation skills in total quality management (TQM) principles and continuous process improvement methods in the manufacturing environment.

The purpose of the *40 Top Tools for Manufacturers* is to provide background information and overviews of tool usage that will assist teams in selecting, sequencing, and applying major TQM tools, methods, and processes such as just-in-time (JIT), total productive maintenance (TPM), process mapping (P-M), cycle time management (CTM), and many others. These major tools and processes are instrumental in any organizational change effort such as business process reengineering, switching to integrated product development teams, or establishing a JIT manufacturing system. Figure I-1 shows a system analysis diagram displaying the application of the major TQM tools, methods, and processes (see Appendix A for additional definitions of terms). Another version of this figure shows a system analysis diagram emphasizing the application of the major manufacturing methods (see Figure I-2). For example, process mapping and cycle time management (CTM) tools can be successfully used to reduce processing time anywhere from the system's supplier side to its customer side, whereas statistical process control (SPC) is typically used more on the product output side, as displayed on the diagram.

In order to make sense of this tool inventory and effectively use this toolbox of 40 manufacturing tools, review the following nine components (see Figure I-3, sample pages):

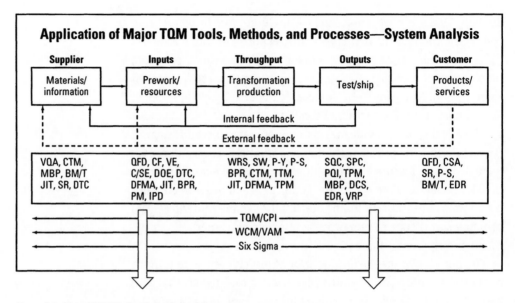

Figure I-1. Major TQM Tools, Methods, and Processes

1. Tool number and name.
2. Tool *also-known-as* (aka).
3. Tool description.
4. Typical application. The classification of the tool suggests the tool's particular process application.
5. Problem-solving phase. Each tool is marked as applicable in one or more of six suggested problem-solving phases.
6. Probable links to other tools in *Tool Navigator*™ (*before* and *after*). Though most of these tools do not appear in this book, the link box still provides the reader with a list of additional tools available to them. (For a complete description of the tools, refer to *Tool Navigator*™.)
7. Notes and key points.

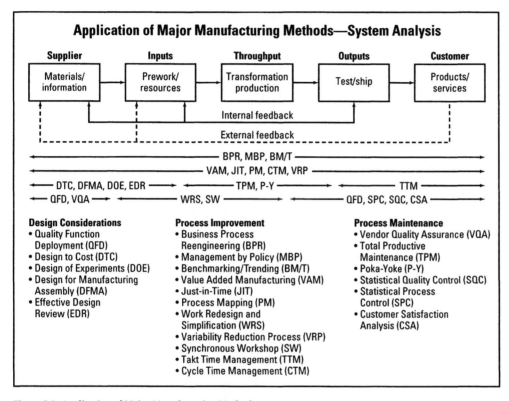

Figure I-2. Application of Major Manufacturing Methods

8. Step-by-step procedure to explain how the team is to use each tool.

9. Example of tool application (the output or result). Realistic source data or a problem/opportunity has been used to produce an expected output in the form of a matrix, sketch, flowchart, diagram, graph, table, map, list, or whatever a particular tool produces.

Figure I-3. Sample Two-Page Spread Detailing the Components of Each Tool and
Their Intended Use.

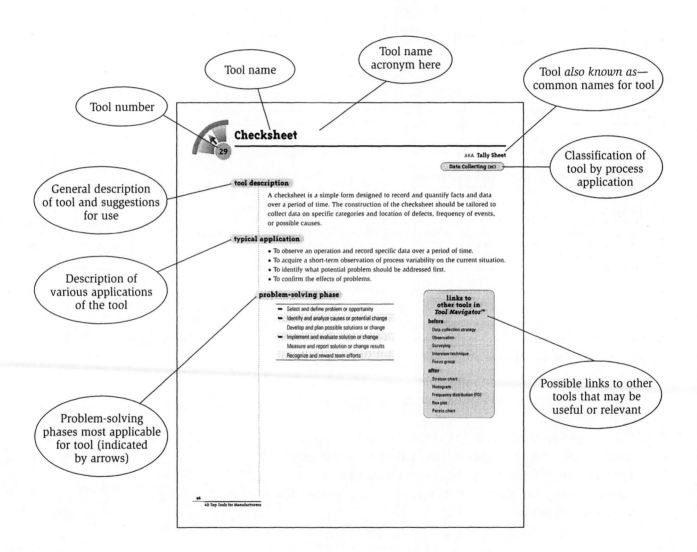

Helpful and supporting information

Step-by-step procedure explaining how to use tool

Example of the display, output, or result of the tool's use

Checksheet · TOOL 29

notes and key points

Types of frequently used checksheets:
- To count occurrences (tally Ⲏ).
- To measure activities (amounts, time, etc.)
- To locate problems or defects (defect map)

step-by-step procedure

STEP 1 Identify data to be collected. See example *Customer Complaints*.

STEP 2 Design checksheet for easy data capture.

STEP 3 Collect data for the stated time period. Example: 20 days.

STEP 4 After a specified period, total the check marks and input this data into the problem-solving process. Date the checksheet.

STEP 5 Note: For another type of checksheet, see the defect map approach.

example of tool application

Customer Complaints   Date: xx/xx/xx

Type	Week 1					Week 2					Week 3					Week 4					Total
	M	T	W	T	F	M	T	W	T	F	M	T	W	T	F	M	T	W	T	F	
Ordering	II	I	Ⲏ	I	III	II	Ⲏ	Ⲏ	III	Ⲏ	III	Ⲏ	I	ⲎI		Ⲏ	III	II	Ⲏ	IIII	66
Shipping	I	II		II	I	I		III			II		I	II	I	I	I		II		20
Billing	I	I			I			II		III	I	I			III	II	I	II			20
Defect	II	I	Ⲏ	Ⲏ	II	I	II	III	I	I	II		II	Ⲏ		I	I	II	I	III	41
Service	Ⲏ	I	III	II	Ⲏ	II	I	Ⲏ	I	Ⲏ	Ⲏ	Ⲏ	II	Ⲏ	I	ⲎI	II	Ⲏ	I	Ⲏ	69
Total	11	6	14	12	11	7	7	18	6	14	13	12	5	18	5	15	9	11	9	13	216

97
40 Top Tools for Manufacturers

I have also sorted and arranged the 40 tools into four manufacturing stages using a generic TQM or continuous improvement approach. The four stages are:

1. manufacturing process
2. cycle time
3. variability
4. problem solving

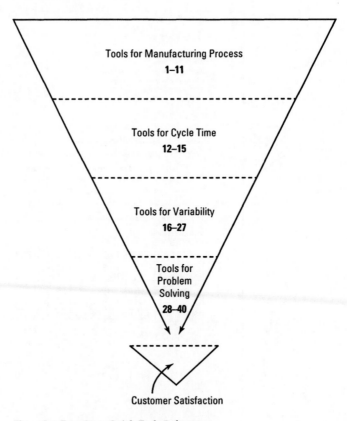

Figure I-4. Four-Stage Quick Tools Reference

The classification of these tools in these manufacturing stages will help teams select the appropriate tools within each stage for problem-solving or manufacturing process improvement activities. The first three stages help you determine and narrow down problematic conditions in your manufacturing process while the fourth stage provides you with the necessary problem-solving techniques to solve problems and improve your process. By following these four stages you will reach customer satisfaction (see Figure I-4).

Though the 40 navigator manufacturing tools are the same tools you will find in the *Tool Navigator™—The Master Guide for Teams*, I have renumbered them sequentially in this book. The 40 tools in this book appear as follows (the number in parentheses is the tool's number in the Master Guide):

Stage One—Navigator Tools for the Manufacturing Process

These 11 navigator manufacturing tools will help you to verify, or suggest further action to improve the process.

1 **(7)**	5 **(146)**	9 **(203)**
2 **(28)**	6 **(149)**	10 **(220)**
3 **(72)**	7 **(150)**	11 **(221)**
4 **(89)**	8 **(151)**	

Stage Two—Navigator Tools for Cycle Time

Once you have determined your manufacturing process you must look at excessive cycle times. These four navigator manufacturing tools will measure and help you to reduce your manufacturing activity cycle time.

12 **(5)**	14 **(111)**
13 **(56)**	15 **(201)**

Stage Three—Navigator Tools for Variability

Once you determine your cycle times these 12 navigator manufacturing tools will help you monitor and identify the variability (specification deviations or defects) in your manufacturing process.

16 (**10**)	20 (**44**)	24 (**114**)
17 (**40**)	21 (**71**)	25 (**147**)
18 (**41**)	22 (**83**)	26 (**191**)
19 (**42**)	23 (**101**)	27 (**212**)

Stage Four—Navigator Tools for Problem-Solving

Now that you have followed the first three steps you are ready to use the following 13 navigator manufacturing tools to help you analyze and perform problem-solving techniques on specific problems or improvement opportunities.

28 (**26**)	33 (**75**)	38 (**174**)
29 (**29**)	34 (**106**)	39 (**183**)
30 (**60**)	35 (**130**)	40 (**218**)
31 (**64**)	36 (**138**)	
32 (**74**)	37 (**145**)	

Included among the 40 selected tools are some tools typically used for areas such as engineering or change management, rather than manufacturing. For example, the analysis of variance (tool 16 in this book) or the cycle time flowchart (tool 13 in this book). The rationale for including some of these tools is that they are powerful and very useful, and they apply to many categories or disciplines besides manufacturing.

Table I-1 will also help the reader easily identify, select, and sequence these 40 tools according to the problem-solving or manufacturing improvement phase. The tools are arranged alphabetically.

As mentioned earlier, most of the *before* and *after* links to other tools are not in this book. Appendix B alphabetically lists the 222 tools (as well as their *aka*'s) as they appear in the *Tool Navigator™— The Master Guide for Teams*. I have highlighted other manufacturing and supporting tools in Appendix B to help the reader link and

Table I-1. Problem-Solving Phases of the 40 Top Manufacturing Improvement Tools (arranged alphabetically)

Classification of Tools

TB	—	Team Building
IG	—	Idea Generating
DC	—	Data Collecting
AT	—	Analyzing/Trending
ES	—	Evaluating/Selecting
DM	—	Decision Making
PP	—	Planning/Presenting
CI	—	Changing/Implementing

Problem-Solving Phases

1. Select and define problem or opportunity
2. Identify and analyze causes or potential change
3. Develop and plan possible solution or change
4. Implement and evaluate solution or change
5. Measure and report solution or change results
6. Recognize and reward team efforts

Tool Number in *40 Top Tools for Manufacturers*	Tool Number in *Tool Navigator*™	Class	Tool Name	P-S Phase Number					
				1	2	3	4	5	6
12	(5)	AT	Activity analysis	•	•				
1	(7)	PP	Activity network diagram				•	•	
16	(10)	AT	Analysis of variance	•	•		•	•	
28	(26)	AT	Cause and effect diagram adding cards (CEDAC)	•	•				
2	(28)	IG	Checklist		•		•		•
29	(29)	DC	Checksheet	•	•	•			
17	(40)	AT	Control chart—c (attribute)	•	•		•	•	
18	(41)	AT	Control chart—p (attribute)	•	•		•	•	
19	(42)	AT	Control chart—X–R (variable)	•	•		•	•	
20	(44)	AT	Cost of quality		•	•			
13	(56)	AT	Cycle time flowchart	•	•				
30	(60)	DC	Defect map	•	•				
31	(64)	ES	Dendrogram	•	•	•			
21	(71)	CI	Events log	•	•	•			
3	(72)	CI	Facility layout diagram	•	•	•			
32	(74)	CI	Failure mode effect analysis (FMEA)	•	•	•			
33	(75)	AT	Fault tree analysis (FTA)	•	•	•			
22	(83)	AT	Frequency distribution	•	•				
4	(89)	PP	Gozinto chart	•	•				
23	(101)	AT	Line chart	•	•		•	•	
34	(106)	PP	Matrix diagram		•		•		
14	(111)	CI	Monthly assessment schedule		•	•	•	•	
24	(114)	AT	Multivariable chart	•	•				
35	(130)	AT	Pareto chart	•	•		•	•	
36	(138)	CI	Potential problem analysis (PPA)	•	•	•			
37	(145)	PP	Problem specification	•	•				
5	(146)	AT	Process analysis	•	•	•			
25	(147)	AT	Process capability ratios	•	•			•	
6	(149)	AT	Process flowchart	•	•				
7	(150)	CI	Process mapping	•	•				

continued

Table I-1. Problem-Solving Phases of the 40 Top Manufacturing Improvement Tools, *continued*

Tool Number in *40 Top Tools for Manufacturers*	Tool Number in *Tool Navigator™*	Class	Tool Name	P-S Phase Number					
				1	2	3	4	5	6
8	(151)	CI	Process selection matrix			•	•		
38	(174)	AT	SCAMPER	•	•	•			
39	(183)	ES	Solution matrix			•		•	•
26	(191)	AT	Stratum chart	•	•		•		
15	(201)	CI	Time study sheet	•	•		•	•	
9	(203)	PP	Top-down flowchart		•	•			
27	(212)	AT	Variance analysis	•	•			•	
40	(218)	AT	Window analysis	•	•	•			
10	(220)	PP	Work breakdown structure (WBS)	•	•	•			
11	(221)	CI	Work flow analysis (WFA)		•	•	•		

cross-reference to other tools required for process and quality improvement. Readers can use information in the "links to other tools in *Tool Navigator™*" box to get an idea of what their next step or tool might be or what tools they may want to use in preparing for the next step. For example, data collection tools typically are used to input data when using control charts (tools 17–19 in this book). On the other hand, cost-benefit analysis tools are performed after the cost of quality tool (tool 44 in the Master Guide) is applied.

Tools do not solve problems, people do. Just as a team facilitator makes it easier for a team to problem solve, the application of the appropriate tools by a team facilitates their problem-solving process. From my 30 years of experience collecting material and notes on tools and techniques, I know that it is important to have the right tool when you need it. I believe the *40 Top Tools for Manufacturers* gives teams the tools necessary to start and complete almost any manufacturing task.

Navigator Tools for
the Manufacturing Process

Activity Network Diagram

AKA **Arrow Analysis, Node Diagram**

Planning/Presenting (PP)

tool description

The activity network diagram is a project planning and scheduling tool for product development or improvements. It graphically displays the sequential flow of activities, estimated time requirement and start/finish times, critical path, and interrelationship of activities.

typical application

- To map and schedule in a logical sequence all required activities in a project to be completed.
- To identify a critical path and resource allocations.
- To coordinate and control parallel activities, estimated completion times, and to meet critical data deadlines.

problem-solving phase

Select and define problem or opportunity

Identify and analyze causes or potential change

Develop and plan possible solutions or change

➥ Implement and evaluate solution or change

➥ Measure and report solution or change results

Recognize and reward team efforts

links to other tools in *Tool Navigator*™

before

Work breakdown structure (WBS)

Top-down flowchart

Activity analysis

Project prioritization matrix

Process analysis

after

Responsibility matrix

Milestones chart

Gantt chart

Project planning log

Resource histogram

notes and key points

Activity		
Task I.D.	ES	EF
Time	LS	LF

ES = Early Start
EF = Early Finish
LS = Late Start
EF = Late Finish

Bold Lines ➡ = Critical path = Represents
The longest completion time from starting
the first task to finishing the last task

Slack (Float):
1a − 1b − 1c = 24 days ∴ 0 days slack (critical path)
2a − 2b − 1c = 20 days ∴ 4 days slack
1a − 1b − 3b = 17 days ∴ 7 days slack
1a − 3a − 3b = 14 days ∴ 10 days slack

step-by-step procedure

STEP 1 Complete a work breakdown structure (WBS) or similar data collection activity to identify and sequence project activities to be completed. See example *Development of a Statistical Process Control (SPC) Training Course*.

STEP 2 Sort and sequence all activities from left to right, determine parallel paths and interrelationship.

STEP 3 Complete, on Post-its, all required information as shown in *notes and key points* and this example.

– Record name of activities, task identification, estimated completion time, and early/late start and finish times. For calculating the critical path, add all estimated times from start to finish of the project. This is also the longest completion time and the earliest time that the project can be completed. There is no slack (float) time.

– For calculating early start/finish times, add the estimated time for each task (left-to-right) to the cumulative duration of the preceding tasks.

– For calculating late start/finish times, subtract the estimated time for each task (right-to-left) from the late start (LS) time of the succeeding tasks.

– For calculating slack or float time, determine the differences (if any) between the early start (ES) and the late start (LS) for each task. Also, calculate slack in each path of the diagram as shown in the example.

step-by-step procedure (continued)

STEP 4 Finalize the diagram by chaining all nodes and checking sequential and logical flow.

STEP 5 Check all information, title and date the chart.

example of tool application

Development of an SPC Training Course   Date: xx/xx/xx

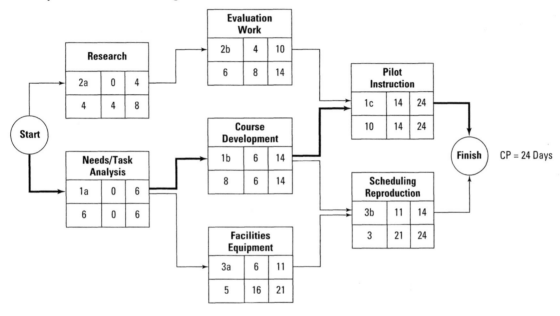

Note: ——► Heavy line = critical path (CP)

Checklist

Idea Generating (IG)

tool description

A checklist is a useful tool for guiding a team's activities and progress, providing important steps and information in a procedure, collecting and organizing data, and helping in the idea generation process for product development and problem solving. Checklists can also be useful as work instructions and safety checks.

typical application

- To prevent the omission of critical steps in a process or procedure.
- To question if certain items or ideas have been completed or considered.
- To collect and organize data for problem analysis.

problem-solving phase

Select and define problem or opportunity

➡ Identify and analyze causes or potential change

Develop and plan possible solutions or change

➡ Implement and evaluate solution or change

Measure and report solution or change results

Recognize and reward team efforts

links to other tools in *Tool Navigator*™

before

Data collection strategy

Observation

Questionnaires

Critical incident

Thematic content analysis

after

SCAMPER

What-if analysis

SWOT analysis

Stimulus analysis

Problem specifications

notes and key points

- Some of the most common checklists cover the following areas:
  - New product development
  - Problem prevention
  - Idea generation for solutions
  - Start-up and progress
  - Selection or prioritization
  - Work instructions
  - Data collection and recording

step-by-step procedure

STEP 1   As a first step, the purpose and intended use of the checklist is determined.

STEP 2   Research is then performed to ensure that the developed checklist covers all requirements, provides all options, or asks for specific data to be recorded. See example *Checklists for Teaming*.

STEP 3   When constructing the checklist, provide space for checking off completed steps, ideas, or data items, as shown in the example.

STEP 4   Ask subject matter experts to review the final draft of a checklist to ensure that nothing of importance has been overlooked or omitted.

STEP 5   Perform final revisions and pilot the checklist.

example of tool application

Checklists for Teaming   Date: xx/xx/xx

✓	Team Start-Up Sequence	✓	Team Norms	✓	Generate Ideas for Solutions
	Organizational readiness?		Start and end on time		Change materials
	Top management support?		No off-side conversations		Change work instructions
	A champion coordinating?		Participate—active contribution		Change color or symbols
	Volunteers for teams?		Assists keeping team focused		Change shape or format
	Schedule and facility ready?		Avoid interrupting others		Change size or amount
	Team training available?		Equal status for all		Change design or style
	Team role assignments made?		No evaluation of team members		Change person or place
	Team norms established?		Allow process flexibility		Rearrange sequence
	Mission and goals developed?		Be open to new ideas		Rearrange parts
	Problem specification stated?		Help facilitate		
	Team meetings scheduled?		Complete assigned action item		

Facility Layout Diagram

3

Changing/Implementing (CI)

tool description

A facility layout diagram displays floor layouts, manufacturing or service process flows, employee movement, possible bottlenecks, excessive cycle time, and other possible inefficiencies in the present layout. This tool prepares a layout of equipment, work areas, and storage areas useful in work flow analysis (WFA) and improvement efforts.

typical application

- To create a floor plan of the facility for tracing employee movement.
- To identify possible co-location of activities and shorter distances of required movement.
- To map process flows and loopbacks.

problem-solving phase

➥ Select and define problem or opportunity

➥ Identify and analyze causes or potential change

➥ Develop and plan possible solutions or change

Implement and evaluate solution or change

Measure and report solution or change results

Recognize and reward team efforts

links to other tools in *Tool Navigator*™

before

Process flowchart

Cycle time flowchart

Opportunity analysis

Mental imaging

Interview technique

after

Work flow analysis (WFA)

Activity analysis

Variance analysis

Potential problem analysis

What-if analysis

notes and key points

For easy reference to any work area, use grid labeling of facility layout. Example:

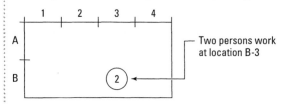

step-by-step procedure

STEP 1   First, identify and observe the facility or work area for the normal work flow, equipment location, and the number of employees normally assigned to workstations. See example *Reproduction Facility Layout*.

STEP 2   Using drafting paper, prepare a facility layout diagram showing all major items or equipment, furniture, and other objects or areas of interest.

STEP 3   Name all drawn items, complete grid reference labeling, and identify the number of people working at certain locations.

STEP 4   Finally, check for accuracy and date the diagram.

example of tool application

Reproduction Facility Layout

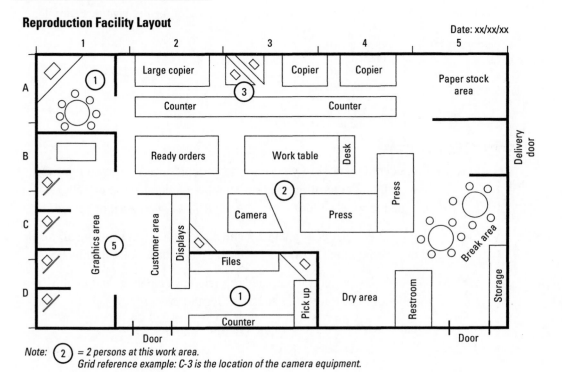

Note: (2) = 2 persons at this work area.
Grid reference example: C-3 is the location of the camera equipment.

Gozinto Chart

Planning/Presenting (PP)

tool description

The gozinto chart is a vertical tree diagram that displays hierarchical levels of detail of a complete product assembly to ship process. Developed by A. Vazsonyi, this project planning tool is of great value for kitting, bill of materials auditing, parts/number identification, and operator training.

typical application

- To breakdown a product into its parts.
- To flow out the assembly process.
- To cross-reference parts data with the hierarchical levels of assembly.

problem-solving phase

➡ Select and define problem or opportunity

➡ Identify and analyze causes or potential change

➡ Develop and plan possible solutions or change

Implement and evaluate solution or change

Measure and report solution or change results

Recognize and reward team efforts

notes and key points

- Gozinto chart numbering is by levels and BOM or part identification number. Example: 3015: 3 for level 3, 015 for bill of material or part ID number.
- This is a similar approach to the work breakdown structure (WBS) chart.

links to other tools in *Tool Navigator*™

before

Tree diagram

Work breakdown structure (WBS)

Information needs analysis

Process analysis

Work flow analysis (WFA)

after

Failure mode and effect analysis

Task analysis

Potential problem analysis

Dendrogram

Activity analysis

step-by-step procedure

STEP 1   List all parts required to completely assemble the product.

STEP 2   Draw a hierarchy of assembly, showing levels of detail from the top down to the basic level of parts.

STEP 3   Provide identification of parts; name and number each part charted.

STEP 4   Check completeness of chart and date.

example of tool application

Mousetrap — Assembly to Ship

Date: xx/xx/xx

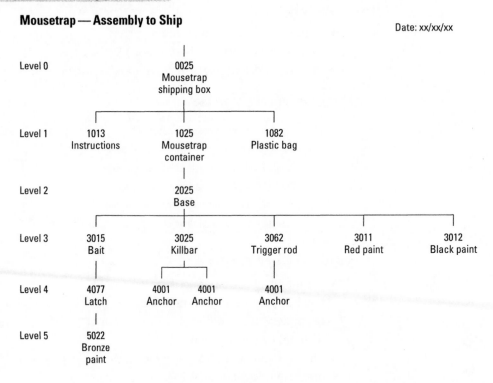

Process Analysis

AKA **N/A**

Analyzing/Trending (AT)

tool description

The process analysis technique helps to trace the source of variation and is, therefore, a useful method to identify root causes of a problem. Process analysis is typically performed using an activity-level process flowchart and by asking a series of questions to explore or justify excessive cycle time, approvals, improper sequence, delays, and other process deficiencies.

typical application

- To review, analyze, and improve an existing process.
- To identify process improvement opportunities.
- To fine-tune processes in an organizational change project.

problem-solving phase

➡ Select and define problem or opportunity

➡ Identify and analyze causes or potential change

➡ Develop and plan possible solutions or change

Implement and evaluate solution or change

Measure and report solution or change results

Recognize and reward team efforts

links to other tools in *Tool Navigator*™

before

Symbolic flowchart

Organization chart

Process mapping

Process flowchart

Cycle time flowchart

after

Activity analysis

Variance analysis

Work flow analysis (WFA)

Facility layout diagram

Decision process flowchart

notes and key points

- To construct a process flow, several tools are available:
  - process flowchart
  - symbolic flowchart
  - process mapping
  - cycle time flowchart
  - activity analysis

  Using any one of these will allow a process improvement team to achieve established team goals.
- The given list of 10 process analysis questions is optional. The number and content of questions may change in accordance with the complexity of any given process.

step-by-step procedure

STEP 1   As a prerequisite activity, a facilitated team develops a process flowchart at the activity-level for the process selected.

STEP 2   A set of standard process analysis questions is displayed by the facilitator. The team reviews the questions, adds, deletes, or revises questions to fully cover the process to be analyzed.

STEP 3   Using the finalized list of questions, the team discusses all activities in the process and provides responses to the questions.

STEP 4   Finally, the facilitator asks participants to recheck all responses, makes final revisions, and dates the list.

STEP 5   The information serves as an input to a variance analysis process, a logical next step for the team.

example of tool application

Symbolic Flowchart for the Facilitation of Process Mapping

Date: xx/xx/xx

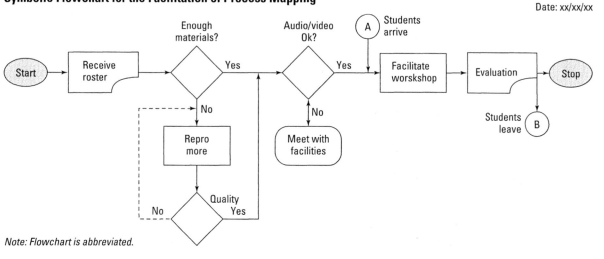

Note: Flowchart is abbreviated.

Typical Process Analysis Questions	Date: xx/xx/xx
1. Are the connected tasks performed in a logical sequence? *No, materials check should have been done earlier*	
2. Does the defined process show more than two loopbacks? *Yes*	
3 Do individual tasks have relatively long cycle times? *No*	
4. Does every task add value to the process? *No, audio-visual check does not add value to this process.*	
5. Are there redundant tasks? *No*	
6. Does the process reflect excessive delays? *No*	
7. Does the process contain sources of key variance? *No*	
8. Are there more than two approval requirements? *No*	
9. Can the process flow be changed to reduce tasks? *Yes, remove materials and A/V checks*	
10. Does this process have a high level of consistency? *Yes*	

Process Flowchart

6

tool description

A process flowchart illustrates the major activities, sequence, and flow connections of a work process or project. The flowchart helps a team gain a common understanding of the overall process and its interrelationships. The flowchart can be used to identify problem areas, document a process, or serve as a planning tool for process improvement.

typical application

- To illustrate the flow or process step in manufacturing a product, providing a service, or managing a project.
- To provide a common understanding of a complex process.
- To recommend an improved process to the process owner.

problem-solving phase

➡ Select and define problem or opportunity

➡ Identify and analyze causes or potential change

Develop and plan possible solutions or change

Implement and evaluate solution or change

Measure and report solution or change results

Recognize and reward team efforts

links to other tools in *Tool Navigator™*

before

Information needs analysis

Systems analysis diagram

Variance analysis

Problem analysis

Pareto chart

after

Process mapping

Problem specification

Opportunity analysis

Work flow analysis (WFA)

Action plan

notes and key points

- Legend for process flow chart symbols:

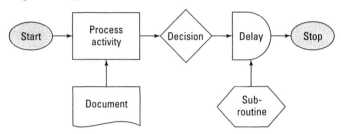

Note: Ensure that participants represent and cover all process areas to be flowcharted.

step-by-step procedure

STEP 1 As a first step, the team facilitator assembles a team whose participants thoroughly understand all aspects of the process. See example *Reengineering Human Resources*.

STEP 2 The overall scope of the process flowchart is determined. A starting and stopping point is identified.

STEP 3 Next, participants identify all major process steps and the sequence of completion. Reviews, delays, documents, reports, and other important activities are recorded on another flip chart.

STEP 4 The facilitator uses a whiteboard to start drawing the process flowchart. The participants assist the facilitator in drawing and connecting all process steps in the correct sequence.

STEP 5 Depending on the level of detail agreed upon, additional information can be noted along with the process steps, as shown in the example.

STEP 6 Finally, the process flowchart is verified for accuracy and dated.

example of tool application

Reengineering Human Resources   Date: xx/xx/xx

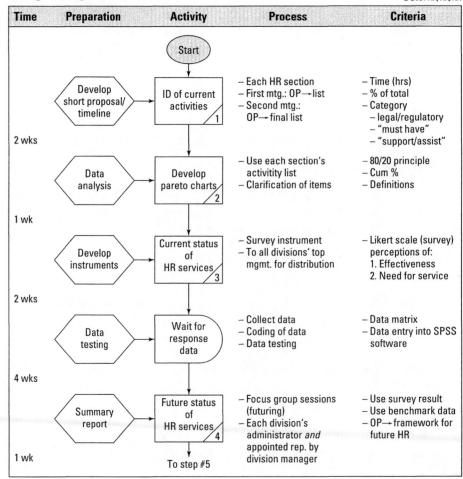

Time	Preparation	Activity	Process	Criteria
		Start		

2 wks	Develop short proposal/ timeline	ID of current activities **1**	– Each HR section
			– First mtg.: OP→list
			– Second mtg.: OP→final list
				– Time (hrs)
				– % of total
				– Category
				  – legal/regulatory
				  – "must have"
				  – "support/assist"

1 wk	Data analysis	Develop pareto charts **2**	– Use each section's activity list – Clarification of items	– 80/20 principle – Cum % – Definitions
2 wks	Develop instruments	Current status of HR services **3**	– Survey instrument – To all divisions' top mgmt. for distribution	– Likert scale (survey) perceptions of: 1. Effectiveness 2. Need for service
4 wks	Data testing	Wait for response data	– Collect data – Coding of data – Data testing	– Data matrix – Data entry into SPSS software
1 wk	Summary report	Future status of HR services **4**	– Focus group sessions (futuring) – Each division's administrator *and* appointed rep. by division manager	– Use survey result – Use benchmark data – OP→framework for future HR
		To step #5		

continued

Reengineering Human Resources, *continued*

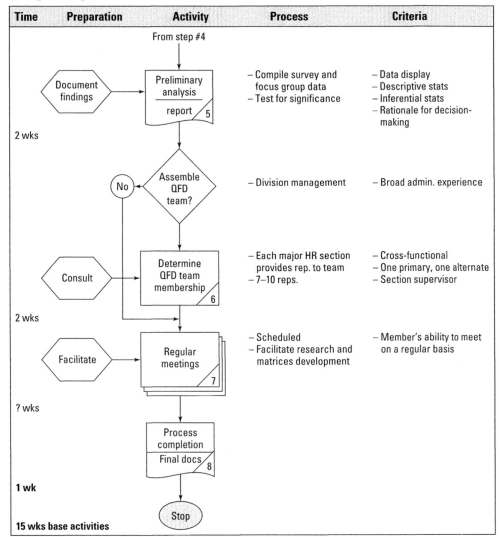

Time	Preparation	Activity	Process	Criteria
		From step #4		
	Document findings	Preliminary analysis — report 5	– Compile survey and focus group data – Test for significance	– Data display – Descriptive stats – Inferential stats – Rationale for decision-making
2 wks		Assemble QFD team? No	– Division management	– Broad admin. experience
	Consult	Determine QFD team membership 6	– Each major HR section provides rep. to team – 7–10 reps.	– Cross-functional – One primary, one alternate – Section supervisor
2 wks				
	Facilitate	Regular meetings 7	– Scheduled – Facilitate research and matrices development	– Member's ability to meet on a regular basis
? wks				
		Process completion — Final docs 8		
1 wk				
15 wks base activities		Stop		

Process Mapping

7

(**Changing/Implementing** (CI))

tool description

The process mapping tool is of great value for teams in documenting the existing process. It identifies and maps all cross-functional processes, process owners (organizations), metrics, and estimated processing time or mapped activities. A finalized process map ensures a thorough understanding of the "as is" process and provides baseline input data for a process improvement team.

typical application

- To mark a visual map of the process in order to perform the analysis necessary for identifying problematic conditions.
- To identify, map, analyze, and prepare *as is* and *should be* process maps.
- To draw a map for process understanding and to discover potential areas for improvement.
- To reduce cycle time of mapped activities.

problem-solving phase

➡ Select and define problem or opportunity
➡ Identify and analyze causes or potential change
Develop and plan possible solutions or change
Implement and evaluate solution or change
Measure and report solution or change results
Recognize and reward team efforts

links to other tools in *Tool Navigator*™

before

Affinity diagram

Systems analysis diagram

Pareto chart

Potential problem analysis (PPA)

Needs analysis

after

Cycle time flowchart

Gap analysis

Force field analysis (FFA)

Barriers-and-aids analysis

Activity analysis

notes and key points

- Symbols and scale:

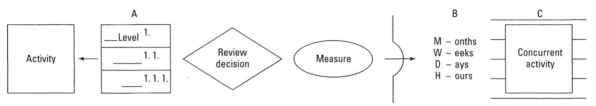

A – Optional level of detail for activities
B – Time scale: Months (M), Weeks (W), Days (D), Hours (H).
C – Four organizations performing activity concurrently

- "Connectors" example:

- A process map can be developed at the macro, mini, or micro level of an organizational process.

step-by-step procedure

STEP 1 A team facilitator assembles a team of cross-functional representatives to assist the development of the process map.

STEP 2 The team decides on the level of detail to be mapped—that is macro (overview), mini (most activities), and micro (very detailed, specific tasks).

STEP 3 Next, the process start and stop points are determined.

STEP 4 As a prerequisite activity, four flip charts are prepared to serve as input data to the process-mapping procedure:

- A listing of all organizations or work groups. Sequence list in order of occurrence.

- A listing of all major functions or activities. Sequence list in order of occurrence.

- A listing of all reviews, audits, approvals, or other decision-making activities. Sequence list in order of occurrence.

step-by-step procedure (continued)

– A listing of all measurements (metrics) in the following categories: process, results, resources, and customer satisfaction. Sequence list in order of occurrence.

STEP 5   The team facilitator, on a whiteboard, draws the process map as directed and checked by the team. The listings of process sequences "organizations," "major functions," "decision-making," and "metrics" are referenced, in order of occurrence, to map out the complete process. See example *Process: Prepare Draft of Action Plans*.

STEP 6   Finally, the team checks the completed map, final revisions are made, and the map is titled and dated. The facilitator redraws the process map on flip charts for future reference.

example of tool application

Process: Prepare draft of action plans   Date: xx/xx/xx
Goal: New process—Improve project management documentation

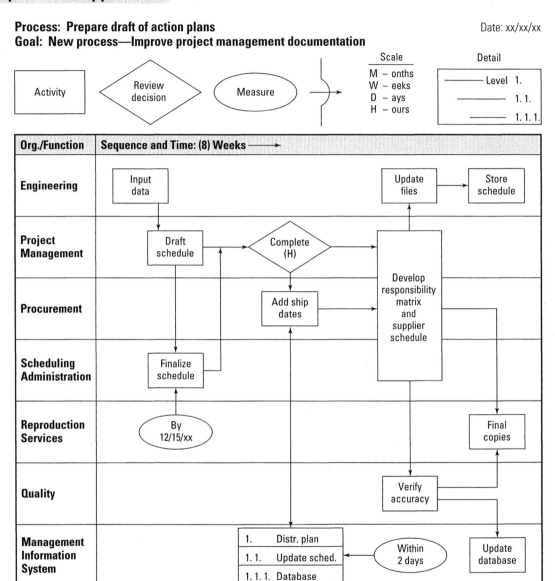

Process Selection Matrix

AKA **N/A**

$\boxed{\text{Changing/Implementing (CI)}}$

tool description

A process selection matrix uses a set of criteria for prioritization to determine a team's first choice. Typically, the team establishes the criteria and rating method. Team consensus is the basis for completing the matrix.

typical application

- To prioritize processes, projects, or systems to be used in a problem-solving or major organizational change effort.
- To identify a process or tool that promises the greatest return on total investment.
- To determine ways to improve the organization.

problem-solving phase

Select and define problem or opportunity

Identify and analyze causes or potential change

➡ Develop and plan possible solutions or change

➡ Implement and evaluate solution or change

Measure and report solution or change results

Recognize and reward team efforts

notes and key points

- If two total scores are tied, add both ranks and divide to assign median to both rank positions. Example: Two scores = 19 for rank position (2) and (3), therefore 2 + 3 = 5/2 = 2.5. Rank 2.5 is assigned to both positions, as shown in the example.

links to other tools in *Tool Navigator*™

before

Cost-benefit analysis

Benchmarking

Consensus decision making

Process analysis

Potential problem analysis (PPA)

after

Activity cost matrix

Action plan

Information needs analysis

Decision process flowchart

Basili data collection method

step-by-step procedure

STEP 1   The team facilitator prepares a selection matrix and lists all previously determined processes. See example *Organizational Change—Process Selection*.

STEP 2   Participants brainstorm a set of criteria to be used in the rating process.

STEP 3   Next, participants rate each process on a scale of *high*, *medium*, *low* using the consensus decision-making technique.

STEP 4   The facilitator records each rating and totals all rows.

STEP 5   Finally, the process selection matrix is dated and Rank 1 (best choice) is circled.

example of tool application

Organizational Change—Process Selection

Process Selection	Customer Impact	Implementation Feasibility	Employee Motivation	Organization's Competitiveness	Return on Investment	Total	Rank
Just-in-time (JIT) manufacturing	H	L	M	M	H	17	4.5
Integrated product development	H	H	H	H	M	23	(1)
Self-managed work teams	M	H	H	M	M	19	2.5
Hoshin planning system	L	M	L	L	L	7	6
ISO-9000 quality system	H	M	M	H	M	19	2.5
Business process reengineering	M	L	H	H	M	17	4.5

Date: xx/xx/xx

Notes: High = 5, medium = 3, low = 1
Ranking: Highest total is best choice, rank (1)

Top-Down Flow Chart

9

Planning/Presenting (PP)

tool description

A top-down flow chart illustrates the major steps in an organizational work process or project. It shows the essential requirements, sequenced from left to right, with a number of substeps listed below each step. This flow chart has an advantage of showing the complete process without too much detail, therefore allowing a team to quickly understand the problem solving or process-improvement opportunities of the process.

typical application

- To display all necessary steps in a work process or project.
- To provide an overall picture of a top-level process.

problem-solving phase

Select and define problem or opportunity

➥ Identify and analyze causes or potential change

➥ Develop and plan possible solutions or change

Implement and evaluate solution or change

Measure and report solution or change results

Recognize and reward team efforts

links to other tools in *Tool Navigator*™

before

Storyboarding

Systems analysis diagram

House of quality

Process selection matrix

Information needs analysis

after

Basili data collection method

Process analysis

Action plan

Resource requirements matrix

Gantt chart

notes and key points

- Designations of Top-Down Flow Chart elements:

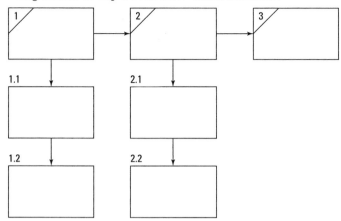

    – Recommendation: A top-down flow chart does not exceed seven process steps (across) and no more than five substeps (top-down).

step-by-step procedure

STEP 1   A facilitator explains the purpose of a top-down flow chart to the team participants. The team identifies the process.

STEP 2   The participants identify essential, major process steps.

STEP 3   Team consensus is reached to select a final 5–7 process steps to be drawn on a whiteboard or flip chart. See example *Motorola's Model of "Six Steps to Six Sigma Quality."*

STEP 4   The facilitator draws the top-down flow chart and asks participants to provide 4–5 substeps for each process step drawn.

STEP 5   Identified substeps are discussed, changed, and finally listed under each major step.

STEP 6   Finally, the facilitator provides numerical identification numbers and sub-step level numbers and dates the flow chart, as shown in the example.

example of tool application

Motorola's Model of "Six Steps to Six Sigma Quality"

Date: xx/xx/xx

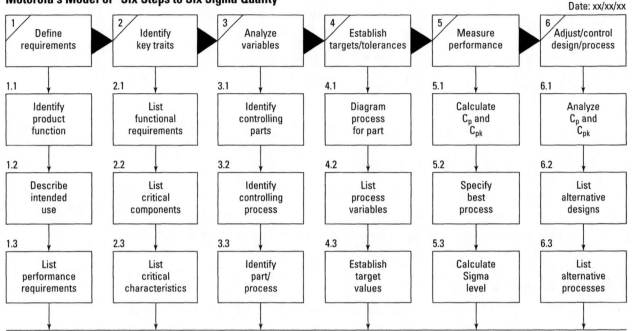

Note: Only a partial flowdown of requirements is shown.

Work Breakdown Structure (WBS)

AKA **Work Breakdown Diagram**

Planning/Presenting (PP)

tool description

A work breakdown structure (WBS) is a necessary division of the overall project into major categories of work. Major categories, in turn, are broken down into more defined, specific elements, and then finally down to a work package level. This process provides project management the ability to schedule, assign resources, and report work package completion status.

typical application

- To break down a total project's work into definable, manageable, and reportable work packages.
- To reduce the complexity of a project so that interrelated activities and work elements can be clearly understood.
- To identify work packages and resource requirements, and to schedule for completing project activities.

problem-solving phase

➥ Select and define problem or opportunity

➥ Identify and analyze causes or potential change

➥ Develop and plan possible solutions or change

Implement and evaluate solution or change

Measure and report solution or change results

Recognize and reward team efforts

links to other tools in *Tool Navigator*™

before

Comparison matrix

Project prioritization matrix

Action plan

Objectives matrix

Responsibility matrix

after

Trend analysis

Gantt chart

Activity network diagram

Program evaluation and review technique (PERT)

Major program status

notes and key points

- A WBS typically consists of five or more levels of breakdown to reduce a project's scope and complexity:

Level	Description	Designation Example
1	Project	10
2	Category	10.1
3	Subcategory	10.1.1
4	Work element	10.1.1.1
5	Work package	10.1.1.1.1
6	Deliverables	10.1.1.1.1.1

step-by-step procedure

STEP 1   The first step for a project manager's team is to identify the major categories of work to be completed. See example *WBS for Adding an Assembly Line*.

STEP 2   A designation or accounting schema is then established to be able to account for or schedule work. The numbering system used is arbitrary—see example shown.

STEP 3   All work categories are broken down into a lower level of detail. This process continues down to the basic work package level. Typically, five or more levels are diagrammed.

STEP 4   The final WBS diagram should reflect all required work and is used as a resource document for the planning and scheduling of the overall project.

example of tool application

WBS for Adding an Assembly Line

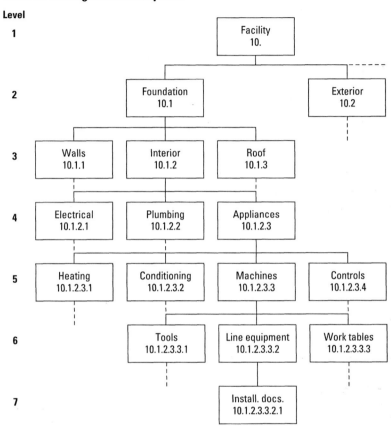

Note: Partial WBS diagram.

Work Flow Analysis (WFA)

AKA **Link Analysis, Assembly Flow**

(**Changing/Implementing** (CI))

tool description

The construction of a work flow analysis (WFA) diagram is absolutely necessary to show how work actually (physically) flows from person to person, via groups and departments throughout the organization. Once completed, the work flow diagram will illustrate process flows, methods, information, and material movement.

Work flow analysis is often used as a problem-solving tool to identify process loops, cross-over, redundant moves, bottlenecks, and other inefficient or non-value-adding activities. It is of great assistance in understanding the existing process before improvements can be proposed in process redesign or reengineering efforts.

typical application

- To show the sequential steps involved in moving people, material, documentation, or information in a process.
- To diagram or baseline work movement in order for a team to understand the current work-flow sequence.
- To illustrate a system's inefficiency.
- To identify and eliminate illogical process flows.

problem-solving phase

Select and define problem or opportunity

➥ Identify and analyze causes or potential change

➥ Develop and plan possible solutions or change

➥ Implement and evaluate solution or change

Measure and report solution or change results

Recognize and reward team efforts

links to other tools in *Tool Navigator*™

before

Facility layout diagram

Fault tree analysis (FTA)

Symbolic flowchart

Gozinto chart

Cycle time flowchart

after

Process analysis

Failure mode effect analysis (FMEA)

Action plan

Decision process flowchart

Consensus decision making

notes and key points

The sequence of flow can be shown by numbering sections of the flow path as shown ————▶. Other process paths can be encoded as – – – – – –▶ or ++++++++++▶.

step-by-step procedure

STEP 1 The team identifies functions and responsibilities of the work-flow process to be analyzed. See example *Picture Frame Assembly.*

STEP 2 The second step is to obtain a current floor plan and to "walk" the flow of the selected process. Mark up the floor plan displaying the basic "as is" flow. For repeats or different process phases, use different color markers. Timing the process at various stages will provide additional data: the cycle times of tasks, movements, and delays. The resulting diagram is sometimes referred to as a "spaghetti diagram" because of the multiline crossover appearance.

STEP 3 The team then performs an analysis of the current process and recommends a first draft of equipment rearrangements and work flow changes to reduce or eliminate process steps.

STEP 4 Step four requires team consensus to be reached and a presentation given to process owners and other interested parties.

STEP 5 Date the completed work flow analysis diagram.

example of tool application

Picture Frame Assembly

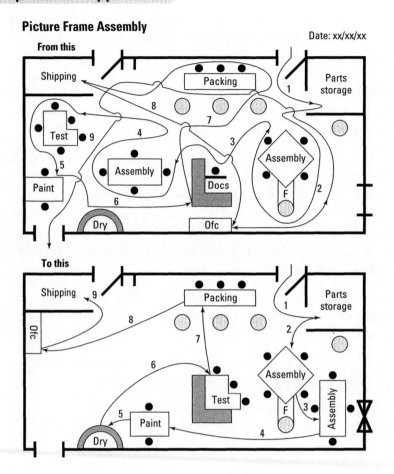

Navigator Tools
for Cycle Time

Activity Analysis

Analyzing/Trending (AT)

tool description

An activity analysis is a very useful tool to account for the time spent in everyday work schedules, programs, or project activities. It allows an analyst to uncover unnecessary work activities, time taken to complete highly important to unimportant activities, and provides data helpful to make work more efficient and effective.

typical application

- To analyze and evaluate work and time requirements to complete the work.
- To identify ineffective work steps and procedures for the purpose of eliminating waste.
- To account for time used within an eight-hour day or other predetermined time period.

problem-solving phase

➡ Select and define problem or opportunity
➡ Identify and analyze causes or potential change
Develop and plan possible solutions or change
Implement and evaluate solution or change
Measure and report solution or change results
Recognize and reward team efforts

links to other tools in *Tool Navigator™*

before

Time study sheet
Task analysis
Checksheet
Check list
Starbursting

after

Activity cost matrix
Value/nonvalue added cycle time chart
Process analysis
Breakdown tree
Cluster analysis

notes and key points

- The following legend pertains to the example *Design Engineer—Two Work Days*.

  Legend: ● = All value-added   Productive activity

                   ◒ = Some value-added   Required activity

                   ○ = Non-value-added   Nonproductive activity

step-by-step procedure

STEP 1   An activity analysis form is prepared for the position, program, or project to be observed and analyzed. See example *Design Engineer—Two Work Days*.

STEP 2   Every activity that is considered important, time-consuming, or required work is timed and recorded on the form.

STEP 3   Other supporting activities are noted for the purpose of time accounting.

STEP 4   All activities are value-rated and coded as an *all, some, non*-value-added activity and recorded as such on the form, as shown in the example.

STEP 5   The totals column is summed to show time used in a specific period of time.

STEP 6   The form is completely filled out in order to provide background data for work simplification or process-improvement efforts.

STEP 7   Finally, the activity analysis form is reviewed and dated.

example of tool application

Design Engineer — Two Work Days

Name:		Date: 11/14/xx	VALUE	Evaluation			Analysis		
Team/Position: Design Engineer							Total		Contact Info.
Date	Time	Activity		Real Work	Routine Work	Busy Work	Hr	Min	
11/12	0745	Get coffee	○						
	0810	Check e-mail	◓		✓			50	
	0900	Return voice mail calls	◓		✓			45	
	0945	Break	○						
	1015	Call program office	●		✓			10	Pgm. J.L.
	1026	Revise drawing AX-221	●	✓			1	35	
	Noon	Lunch							
	1315	Revise drawing AX-255	●	✓			2	25	
	1540	Repro/graphics call	◓		✓			10	8-4445
	1600	Attend IPD meeting	○			✓		45	Weekly
	1650	Call repro	○		✓				8-4445
11/13	0810	Get coffee, talk to boss	○						
	0900	Talk to J.M. to coordinate	◓			✓		30	J.M.
	0930	Check e-mail	◓		✓			30	
	1000	Check voice mail, return calls	◓		✓			30	
	1030	Make calls on design changes	●	✓			1		Boston
	1130	Lunch meeting	○			✓			
	1315	Computer downtime	○						
	1330	Call software group	◓		✓			5	MIS-3
	1410	Supply parts data to G.K.	●	✓				20	G.K.
	1630	Leave for evening class	○						

Note: ● All value-added   ◓ Some value-added   ○ Non-value-added   Total 9 35

Cycle Time Flowchart

AKA **Process Cycle Time Analysis**

Analyzing/Trending (AT)

tool description

A cycle time flowchart accounts for all activities and time required from the start point to the stop point of a process. The intent of using this tool is to identify non-value-adding activities, bottlenecks, excessive loops, approvals, and delays. The flowchart is constructed by a team that owns the process. Participants have a good understanding of the activities and therefore are best suited to collectively produce the cycle time of the overall process.

typical application

- To identify non-value-adding activities and excessive delay time in a process.
- To capture the "as is" process flow in order to have better team understanding of the process for further problem-solving efforts.
- To draw a detailed flowchart for the purpose of completing redesign, problem resolution, and cycle time reduction activities.

problem-solving phase

➡ Select and define problem or opportunity
➡ Identify and analyze causes or potential change
Develop and plan possible solutions or change
Implement and evaluate solution or change
Measure and report solution or change results
Recognize and reward team efforts

links to other tools in Tool Navigator™

before
Process mapping
Problem specification
Pareto chart
Potential problem analysis (PPA)
Systems analysis diagram

after
Process analysis
Problem analysis
Activity analysis
What-if analysis
Force field analysis (FFA)

notes and key points

Symbols and scale:

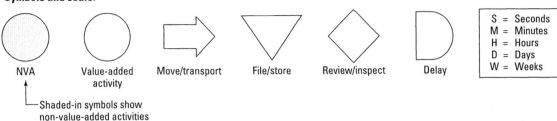

NVA   Value-added   Move/transport   File/store   Review/inspect   Delay
                activity

$\uparrow$
Shaded-in symbols show
non-value-added activities

S = Seconds
M = Minutes
H = Hours
D = Days
W = Weeks

Connectors example:

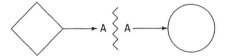

When stating an activity's cycle time, indicate (A) for *actual time*, (E) for *estimated time*.

step-by-step procedure

STEP 1   The team facilitator assembles a team that consists of representatives (process owners) of the process to be the subject of the cycle time flowchart. A whiteboard is prepared to draft out the flowchart.

STEP 2   The team determines the start and stop points of the process. The appropriate scale of time to be used is also determined.

STEP 3   Using team-provided input, the team facilitator, using flowchart symbols, sequentially connects all activities in the process, showing sequence number, what is being done, who does it, and how much time is used to complete each activity. Time is stated as actual (A) or estimated (E). See example *Department Budgeting Process*.

STEP 4   Next, the team checks the charts for completeness, correct task sequence, and task-identifying information.

STEP 5   Finally, the completed cycle time flowchart on the whiteboard is copied onto the flowchart template and the information is summarized. The chart is dated and kept for cycle time reduction work.

example of tool application

Department Budgeting Process

Date: xx/xx/xx

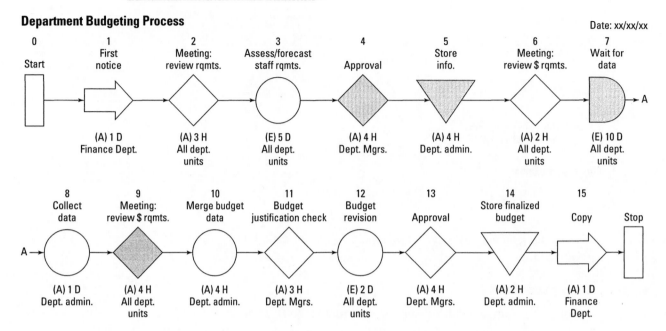

0	1	2	3	4	5	6	7
Start	First notice	Meeting: review rqmts.	Assess/forecast staff rqmts.	Approval	Store info.	Meeting: review $ rqmts.	Wait for data
	(A) 1 D Finance Dept.	(A) 3 H All dept. units	(E) 5 D All dept. units	(A) 4 H Dept. Mgrs.	(A) 4 H Dept. admin.	(A) 2 H All dept. units	(E) 10 D All dept. units

A →

8	9	10	11	12	13	14	15
Collect data	Meeting: review $ rqmts.	Merge budget data	Budget justification check	Budget revision	Approval	Store finalized budget	Copy	Stop
(A) 1 D Dept. admin.	(A) 4 H All dept. units	(A) 4 H Dept. admin.	(A) 3 H Dept. Mgrs.	(E) 2 D All dept. units	(A) 4 H Dept. Mgrs.	(A) 2 H Dept. admin.	(A) 1 D Finance Dept.	

example of tool application   (continued)

Cycle Time Flowchart for Department Budgeting Process

Seq. no.	Cycle time		Symbol	Activity description
	Est.	Act.		
1		8	○ ◇ ⇨ D ▽	First notice
2		3	○ ◇ ⇨ D ▽	Meeting: review rqmts.
3	40		○ ◇ ⇨ D ▽	Assess/forecast staff rqmts.
4		4	○ ◇ ⇨ D ▽	Approval
5		4	○ ◇ ⇨ D ▽	Store info
6		2	○ ◇ ⇨ D ▽	Meeting: review dollar rqmts.
7	80		○ ◇ ⇨ D ▽	Wait for data
8		8	○ ◇ ⇨ D ▽	Collect data
9		4	○ ◇ ⇨ D ▽	Meeting: review dollar rqmts.
10		4	○ ◇ ⇨ D ▽	Merge budget data
11		3	○ ◇ ⇨ D ▽	Budget justification check
12	16		○ ◇ ⇨ D ▽	Budget revision
13		4	○ ◇ ⇨ D ▽	Approval
14		2	○ ◇ ⇨ D ▽	Store finalized budget
15		8	○ ◇ ⇨ D ▽	Copy
16	–	–	○ ◇ ⇨ D ▽	Stop
17			○ ◇ ⇨ D ▽	
18			○ ◇ ⇨ D ▽	
19			○ ◇ ⇨ D ▽	
20			○ ◇ ⇨ D ▽	
21			○ ◇ ⇨ D ▽	
22			○ ◇ ⇨ D ▽	
23			○ ◇ ⇨ D ▽	
24			○ ◇ ⇨ D ▽	
25			○ ◇ ⇨ D ▽	
26			○ ◇ ⇨ D ▽	
27			○ ◇ ⇨ D ▽	
28			○ ◇ ⇨ D ▽	
29			○ ◇ ⇨ D ▽	
30			○ ◇ ⇨ D ▽	

Value added / Non-value added				
	56	12	○ Activity	Time scale: ☐ Seconds
		20	◇ Review/inspect	☐ Minutes
		16	⇨ Move/transport	☒ Hours
	80		D Delay	☐ Days
		6	▽ File/store	☐ Weeks
○○	136	54	Total time: 190 hours	Date: xx/xx/xx

Monthly Assessment Schedule

14

AKA **N/A**

Changing/Implementing (CI)

tool description

A monthly assessment schedule displays all sequential activities of a project, program, or some scheduled event. It reflects the milestone start and completion dates, the length of scheduled activities, and early completion or slipped dates. It serves as a valuable attachment for any status report.

typical application

- To assess monthly the status of scheduled activities for the purpose of overall project management.
- To plan and schedule a project, program or major event.
- To monitor the progress of a project.

problem-solving phase

Select and define problem or opportunity

➥ Identify and analyze causes or potential change

➥ Develop and plan possible solutions or change

➥ Implement and evaluate solution or change

➥ Measure and report solution or change results

Recognize and reward team efforts

links to other tools in *Tool Navigator*™

before

Gap analysis

Force field analysis (FFA)

Barriers-and-aids analysis

Activity analysis

Checklist

after

Action plan

Consensus decision making

Measurement matrix

Countermeasures matrix

Presentation

notes and key points

Schedule milestone dates description:

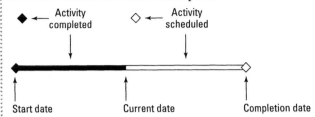

step-by-step procedure

STEP 1   A team assembles for the purpose of identifying and placing into sequential order all of the activities to complete a project, program, or major event. See example *Annual Corporatewide TQM Symposium Plan.*

STEP 2   A monthly assessment schedule form is used to list all activities.

STEP 3   Next, the team reaches consensus on the length of the listed activities and the start and completion dates of each activity.

STEP 4   The draft of the schedule is completed by inserting all milestone dated and connecting lines under the schedule covering a specific time period, as shown in this example.

STEP 5   The team verifies all information, dates the chart, and attaches the chart to an action plan.

example of tool application

Annual Corporatewide TQM Symposium Plan

Date: xx/xx/xx

TQM Symposium Team Schedule

Scheduler: J.K. Smith

Symposium Day: June 19, 199x

#	Activity	Jan 3	10	17	24	31	Feb 7	14	21	28	Mar 7	14	21	28	Apr 4	11	18	25
1.	Select site location	◀━	━	━	━	━	━	━▶										
2.	Negotiate arrangements						◀━	━	━	━▶								
3.	Announce symposium									◀━▶								
4.	Identify dept. reps.			◆														
5.	Call for papers						◆											
6.	Establish guidelines					◀━▶												
7.	Collect papers										◆							
8.	Evaluate papers										◀━	━	◇					
9.	Select papers												◇					
10.	Notify selectees													◇				
11.	Finalize papers															◇		
12.	Establish agenda				◀━	━	━	━	━	━	━	━	━	━	━	◇		
13.	Format symposium									◀━	━	━	━	━	━	━	━	◇
14.	Select speakers											◇━	━	◇				
15.	Concurrent sessions											◇━	━	◇				
16.	Develop rqmts. list											◇◇						
17.	Appoint coordinator											◇						
18.	Develop hosting plan														◇			
19.	Check budget/admin.				◆				◆				◇					◇
20.	Awards and recognition												◇━	━	◇			
21.	Design proceedings								◀━	━	━	━	━	━	━	◇		
22.	– Complete graphics																◇	◇
23.	– Finalize master																	◇
24.	– Repro copies																	
25.	Speaker dry runs																	◇
26.	Final status check																	
27.	Symposium set-up																	
28.	Symposium day																	

Note: For activities #18 → 32 completion dates, see page 2 of this schedule.   ↑ Current date

Time Study Sheet

15

Changing/Implementing (CI)

tool description

The time study sheet is a recording form that displays the cycle time used to complete a task on the first attempt, second attempt, and so on. It can potentially identify a trend or pattern caused by some process variation or change in the use of tools, layout, or sequence steps. The form is also used to show calculations such as average cycle time, various adjustments, and recommended time to be recorded as a standard.

typical application

- To establish criteria of performance.
- To calculate average cycle times for performing tasks.
- To collect cycle time data for constructing a cycle time flowchart.

problem-solving phase

➥ Select and define problem or opportunity

➥ Identify and analyze causes or potential change

   Develop and plan possible solutions or change

➥ Implement and evaluate solution or change

➥ Measure and report solution or change results

   Recognize and reward team efforts

notes and key points

- Record cycle time in seconds, minutes, or hours.
- Make adjustments for performing difficult task elements.
- Make allowances for variations in working conditions.

links to other tools in *Tool Navigator*™

before

Data collection strategy

Observation

Activity analysis

Checksheet

Task analysis

after

Process analysis

Problem specification

Potential problem analysis (PPA)

Variance analysis

Cycle time flow chart

step-by-step procedure

STEP 1 Identify the tasks and elements to be measured. List them on the time study sheet. See example *Calculating Average Cycle Time for Spring Assembly.*

STEP 2 Determine how often elements of the task are to be measured and take measurements. Record measurements and calculate average cycle time ($\bar{X}$CT).

STEP 3 Note any difficulties; make allowances as deemed appropriate.

STEP 4 Summarize and date the time study sheet as shown in this example.

example of tool application

Calculating Average Cycle Time for Spring Assembly

Time Study Sheet (Recording Cycle Time)											Date xx/xx/xx
Task ID#	Element Name and ID#	Cycles—Repeated Measures								$\bar{X}$	Observations on Difficulties and Allowances
		1	2	3	4	5	6	7	8		
1	1A. Attach washer, thread nut	8	10	12	9	12	10			10.2	
	1B. Tighten screw to base	10	12	15	16	11	18			13.7	
	1C. Tension spring to spec	25	35	40	28	32	51			35.2	Often difficult to tension
	1D. Measure spring tension	8	12	7	15	9	8			9.8	
2											

Task ID# 1: Total cycle time per task/elements   413

Number of cycle time measures   24

☒ Seconds ☐ Minutes ☐ Hours $\bar{X}$ CT: 69

Difficulty adjustment: 10

Allowance adjustment: 0

Allowed time: 79

Notes: – $\bar{X}$CT = average cycle time (413 ÷ 6 = 68.8).
– Recheck allowed time periodically.

Navigator Tools
for Variability

Analysis of Variance

16

AKA **ANOVA, Hypothesis Testing (ANOVA), F-Test**

(**Analyzing/Trending (AT)**)

tool description

The analysis of variance is an inferential statistical technique designed to test for significance of the differences among two or more sample means. Some applications include the ability to make inferences about the population from which the samples were drawn, to identify differences or variations in statistical process control (SPC) analyses, and to provide for analysis and comparison of factorial designs in design of experiments (DOE).

typical application

- To identify differences or variance in productivity, quality, methods, factorial designs, performance, and many other applications.
- To check for variation among sample or group means.
- To perform hypothesis testing on interval (quantitative) data sets.

problem-solving phase

➡ Select and define problem or opportunity
➡ Identify and analyze causes or potential change
    Develop and plan possible solutions or change
➡ Implement and evaluate solution or change
➡ Measure and report solution or change results
    Recognize and reward team efforts

links to other tools in *Tool Navigator™*

before

Variance analysis
Standard deviation
Process capability ratios
Normal probability distribution
Descriptive statistics

after

Problem specification
Work flow analysis (WFA)
Prioritization matrix
Process analysis
Problem analysis

notes and key points

- Definition: The analysis of variance (ANOVA) is a technique often applied in the field of inferential statistics to test whether the means of more than two quantative data sets of samples or populations differ.
- Variance analysis:

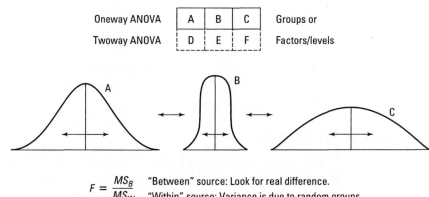

| Oneway ANOVA | A | B | C | Groups or |
| Twoway ANOVA | D | E | F | Factors/levels |

$$F = \frac{MS_B}{MS_W}$$

"Between" source: Look for real difference.
"Within" source: Variance is due to random groups.

- The analysis of variance test is frequently used in testing hypotheses. (See tool 92 Hypothesis Testing in *Tool Navigator*™).
- Partial analysis of variance (ANOVA) distribution table—critical values at the .05 level of significance. (Refer to Appendix—Table C in *Tool Navigator*™ for complete ANOVA critical values table.)

$K - 1 = 3 - 1 = 2$

$N - K = 15 - 3 = 12$

Degrees of freedom for denominator

Degrees of freedom for numerator

	1	2	3	4	5	6	7	8	9	10	12	15	20	24	30	40	60	120	∞
1	161	200	216	225	230	234	237	239	241	242	244	246	248	249	250	251	252	253	254
2	18.5	19.0	19.2	19.2	19.3	19.3	19.4	19.4	19.4	19.4	19.4	19.4	19.4	19.5	19.5	19.5	19.5	19.5	19.5
3	10.1	9.55	9.28	9.12	9.01	8.94	8.89	8.85	8.81	8.79	8.74	8.70	8.66	8.64	8.62	8.59	8.57	8.55	8.53
4	7.71	6.94	6.59	6.39	6.26	6.16	6.09	6.04	6.00	5.96	5.91	5.86	5.80	5.77	5.75	5.72	5.69	5.66	5.63
5	6.61	5.79	5.41	5.19	5.05	4.95	4.88	4.82	4.77	4.74	4.68	4.62	4.56	4.53	4.50	4.46	4.43	4.40	4.37
6	5.99	5.14	4.76	4.53	4.39	4.28	4.21	4.15	4.10	4.06	4.00	3.94	3.87	3.84	3.81	3.77	3.74	3.70	3.67
7	5.59	4.74	4.35	4.12	3.97	3.87	3.79	3.73	3.68	3.64	3.57	3.51	3.44	3.41	3.38	3.34	3.30	3.27	3.23
8	5.32	4.46		3.84	3.69	3.58	3.50	3.44	3.39	3.35	3.28	3.22	3.15	3.12	3.08	3.04	3.01	2.97	2.93
9	5.12	4.26		3.63	3.48	3.37	3.29	3.23	3.18	3.14	3.07	3.01	2.94	2.90	2.86	2.83	2.79	2.75	2.71
10	4.96	4.10	4.07	3.48	3.33	3.22	3.14	3.07	3.02	2.98	2.91	2.85	2.77	2.74	2.70	2.66	2.62	2.58	2.54
			3.86																
11	4.84	3.98	3.71	3.36	3.20	3.09	3.01	2.95	2.90	2.85	2.79	2.72	2.65	2.61	2.57	2.53	2.49	2.45	2.46
12	4.75	3.89		3.26	3.11	3.00	2.91	2.85	2.80	2.75	2.69	2.62	2.54	2.51	2.47	2.43	2.38	2.34	
13	4.67	3.81	3.59	3.18	3.03	2.92	2.83	2.77	2.71	2.67	2.60	2.53	2.46	2.42	2.38	2.34			
14	4.60	3.74	3.49	3.11															
15	4.54																		

step-by-step procedure

STEP 1   First, daily product defect rates are collected on three different methods of production. See example *Three Production Methods and Their Daily Product Defect Rates*.

STEP 2   The Null Hypothesis (H_0) is stated: There is no statistically significant difference in the daily product defect rates and the production methods used measured at .05 alpha (level of significance) using a *F*-test (ANOVA).

STEP 3   The eight-step hypothesis testing procedure is used to arrive at a decision (see Hypothesis Testing [CHI-square] for example).

STEP 4   The calculations are performed in this example.
          Note: The analysis of variance calculations are time consuming and often difficult to calculate. Any basic software program on statistics will perform calculations and provide a printout similar to that shown in this example.

STEP 5   Finally, the test result is verified against the critical value located in the ANOVA distribution table (*notes and key points*). On the basis of the test result *F*-ratio = 5.74 and the critical value = 3.89 (which is lower), the Null Hypothesis (H_0) is rejected. There is a statistically significant difference in the three production methods and their daily product defect rates.

example of tool application

Three Production Methods and Their Daily Product Defect Rates

Step 1 Calculate method totals and means	Step 2 Calculate total sum of squared deviations from the grand mean $\overline{\overline{X}}$	Step 3 Calculate how much scores vary within group from $\overline{X}$ of group
Method A	$X - \overline{\overline{X}} = x \qquad x^2$	$X - \overline{X} = x \qquad x^2$
12	$12 - 7.7 = 4.3 = 18.5$	$12 - 10 = 2^2 = 4$
11	$11 - 7.7 = 3.3 = 10.9$	$11 - 10 = 1^2 = 1$
11	$11 - 7.7 = 3.3 = 10.9$	$11 - 10 = 1^2 = 1$
9	$9 - 7.7 = 1.3 = 1.7$	$9 - 10 = 1^2 = 1$
7	$7 - 7.7 = 0.7 = 0.5$	$7 - 10 = 3^2 = 9$
Σ 50	$\overline{42.5}$	$\overline{16}$
$\overline{X} = 10$		
Method B		
10	$10 - 7.7 = 2.3 = 5.3$	$10 - 7 = 3^2 = 9$
4	$4 - 7.7 = 3.7 = 13.7$	$4 - 7 = 3^2 = 9$
6	$6 - 7.7 = 1.7 = 2.9$	$6 - 7 = 1^2 = 1$
8	$8 - 7.7 = 0.3 = 0.1$	$8 - 7 = 1^2 = 1$
7	$7 - 7.7 = 0.7 = 0.5$	$7 - 7 = 0^2 = 0$
Σ 35	$\overline{22.5}$	$\overline{20}$
$\overline{X} = 7$		
Method C		
4	$4 - 7.7 = 3.7 = 13.7$	$4 - 6 = 2^2 = 4$
5	$5 - 7.7 = 2.7 = 7.3$	$5 - 6 = 1^2 = 1$
7	$7 - 7.7 = 0.7 = 0.5$	$7 - 6 = 1^2 = 1$
6	$6 - 7.7 = 1.7 = 2.9$	$6 - 6 = 0^2 = 0$
8	$8 - 7.7 = 0.3 = 0.1$	$8 - 6 = 2^2 = 4$
Σ 30	$\overline{24.5}$	$\overline{10}$
$\overline{X} = 6$	$SS_t = 89.5$	$SS_w = 46$
Grand total Σ 115 **Grand mean** $\overline{\overline{X}} = 7.7$	$SS_t =$ the sum of all squared deviations of the individual scores from the grand mean of all scores	$SS_w =$ the sum of squared deviations of all the scores in each group from the mean of that group

continued

Three Production Methods and Their Daily Product Defect Rates, *continued*

Step 4

Calculate the sum of all squared deviations of the group means from the grand mean times the number of scores per group (between)!

Method	$\bar{X} - \bar{\bar{X}} = x$	$x^2 \times n$
A	$10 - 7.7 = 2.3$	$5.3 \times 5 = 26.5$
B	$7 - 7.7 = 0.7$	$0.5 \times 5 = 2.5$
C	$6 - 7.7 = 1.7$	$2.9 \times 5 = 14.5$
		$SS_B = 43.5$

Step 5

Calculate degrees of freedom (df):

df for $SS_t = N - 1$	$15 - 1 = 14$	
df for $SS_B = K - 1$	$3 - 1 = 2$	← Numerator
df for $SS_w = N - K$	$15 - 3 = 12$	← Denominator
$SS_t = SS_B + SS_w$	$89.5 = 43.5 + 46$	

Step 6

Calculate the means squares:

$$MS_B = \frac{SS_B}{K\text{-}1} = \frac{43.5}{2} = 21.8$$

$$MS_w = \frac{SS_w}{N\text{-}K} = \frac{46}{12} = 3.8$$

Step 7

Calculate the F-ratio:

$$F = \frac{MS_B}{MS_w} = \frac{21.8}{3.8} = 5.74$$

Step 8

Determine if statistically significant:

F-ratio table at 2 df ⟶ and 12 df ↓ at $\alpha = .05 = 3.89$, $\therefore$ we reject H_0

Step 9

A typical printout table:

Source of variation	Sum of Squares	df	Mean squares	F-ratio	F-probability $\alpha = .05$
Between-groups	43.5	2	21.8	5.74	3.89
Within-groups	46.0	12	3.8		
Total	89.5	14			

Control Chart—c (Variable)

tool description

A control chart is a graph that plots randomly selected data over time in order to determine if a process is performing to requirements or is, therefore, under statistical control. The chart displays whether a problem is caused by an unusual or special cause (correctable error) or is due to chance causes (natural variation) alone.

typical application

- To determine if a process is performing to upper and lower control-limit requirements (process is kept in control).
- To monitor process variations over time, with regard to both special or chance causes.
- To identify opportunities for improving quality and to measure process improvement.
- To serve as a quality measurement technique.

problem-solving phase

➡ Select and define problem or opportunity
➡ Identify and analyze causes or potential change
  Develop and plan possible solutions or change
➡ Implement and evaluate solution or change
➡ Measure and report solution or change results
  Recognize and reward team efforts

links to other tools in *Tool Navigator*™

before

Standard deviation

Sampling methods

Observation

Checksheet

Events log

after

Process capability ratios

Variance analysis

Descriptive statistics

Process analysis

Work flow analysis (WFA)

notes and key points

Types of Control Charts	
Data Required	**For Specific Chart**
Quantitative Variable Data Continuous or measurements *Example:* size, downtime, dimensions, activities per day, etc.	• $\bar{X} - R$ chart[†] (average $\bar{X}$ and range "*R*" of samples) • $\bar{X} - S$ chart (average $\bar{X}$ and standard deviations "*S*" of samples)
Qualitative Attribute Data Discrete or counts *Example:* Complaints, rework, missed due dates, delays, rejects, etc.	• *c* chart[‡] (number of defects in a subgroup) • *np* chart (number of defective units in a subgroup) • *p* chart[†‡] (percentage defective) • μ chart (defects per unit)

Most commonly used charts:

   [†]For variable data: $\bar{X}$-*R* Chart

   [‡]For attribute data: *c* Chart

   [†‡]For attribute data: *p* Chart

Note: For a description of other charts refer to a reference on statistical process control (SPC).

– *c* Chart (attribute data)

– Sample data: Minumum (25) samples, subgroups must be of equal size (sample size is constant).

– Calculations: See *c* Chart example.

$$\bar{c} \ (Avg.) \ = \frac{\text{Total number of defects}}{\text{Number of samples}}, \bar{c} = \frac{350}{25}, \bar{c} = 14$$

Upper Control Limit:

   $UCL = \bar{c} + 3\sqrt{\bar{c}}, UCL = 14 + 3\sqrt{14}, UCL = 25.22$

Lower Control Limit:

   $LCL = \bar{c} - 3\sqrt{\bar{c}}, LCL = 14 - 3\sqrt{14}, LCL = 2.78$

step-by-step procedure

STEP 1 Determine the type of attribute control chart to be used. See example *Typing: Errors per Page* (attribute control chart—Type *c*).

STEP 2 Collect at least 25 samples of data; subgroups must be of equal size.

STEP 3 Prepare a type *c* chart and continue to record collected data as shown. See example chart.

STEP 4 After all 25 subgroups (samples) have been recorded, perform all required calculations. See *notes and key points* above for example.

STEP 5 Plot and connect plotted points to form a trendline. Verify that the trendline points reflect recorded averages ($\bar{c}$).

STEP 6 Analyze plotted data for significant variance or patterns. Date the chart.

example of tool application

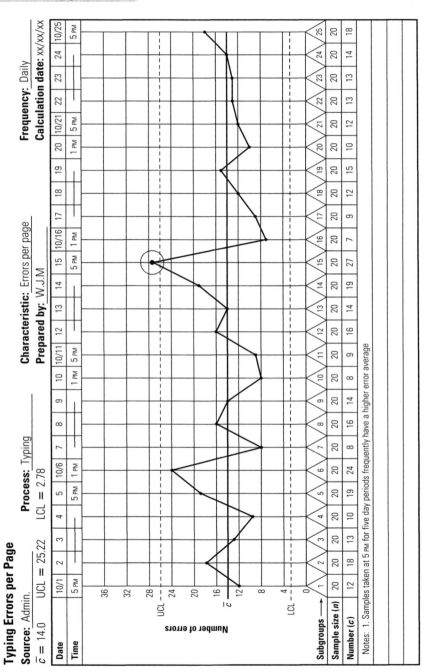

Typing Errors per Page

Source: Admin.   **Process:** Typing   **Characteristic:** Errors per page   **Frequency:** Daily

$\bar{c} = 14.0$ UCL = 25.22   LCL = 2.78   **Prepared by:** W.J.M   **Calculation date:** xx/xx/xx

Date	10/1	2	3	4	5	10/6	7	8	9	10	10/11	12	13	14	15	10/16	17	18	19	20	10/21	22	23	24	10/25
Time	5 PM				5 PM	1 PM				1 PM	5 PM				5 PM	1 PM				1 PM	5 PM				5 PM

Subgroups →	1	2	3	4	5	6	7	8	9	10	11	12	13	14	15	16	17	18	19	20	21	22	23	24	25
Sample size (*n*)	20	20	20	20	20	20	20	20	20	20	20	20	20	20	20	20	20	20	20	20	20	20	20	20	20
Number (*c*)	12	18	13	10	19	24	8	16	14	8	9	16	14	19	27	7	9	12	15	10	12	13	13	14	18

Notes: 1. Samples taken at 5 PM for five day periods frequently have a higher error average

Control Chart—*p* (Attribute)

18

Analyzing/Trending (AT)

tool description

A control chart is a graph that plots randomly selected data over time in order to determine if a process is performing to requirements and is, therefore, under statistical control. The chart displays whether a problem is caused by an unusual or special cause (correctable error) or is due to chance causes (natural variation) alone.

typical application

- To determine if a process is performing to upper and lower control-limit requirements (process is kept in control).
- To monitor process variations over time, with regard to both special or chance causes.
- To identify opportunities for improving quality and to measure process improvement.
- To serve as a quality measurement technique.

problem-solving phase

➤ Select and define problem or opportunity

➤ Identify and analyze causes or potential change

Develop and plan possible solutions or change

➤ Implement and evaluate solution or change

➤ Measure and report solution or change results

Recognize and reward team efforts

links to other tools in *Tool Navigator™*

before

Variance analysis

Sampling methods

Observation

Checksheet

Events log

after

Process capability ratios

Standard deviation

Descriptive statistics

Process analysis

Work flow analysis (WFA)

notes and key points

Types of Control Charts	
Data Required	**For Specific Chart**
Quantitative Variable Data Continuous or measurements *Example:* size, downtime, dimensions, activities per day, etc.	• $\bar{X} - R$ chart[†] (average $\bar{X}$ and range "R" of samples) • $\bar{X} - S$ chart (average $\bar{X}$ and standard deviations "S" of samples)
Qualitative Attribute Data Discrete or counts *Example:* Complaints, rework, missed due dates, delays, rejects, etc.	• c chart[‡] (number of defects in a subgroup) • np chart (number of defective units in a subgroup) • p chart[††] (percentage defective) • μ chart (defects per unit)

Most commonly used charts:

[†]For variable data: $\bar{X}$-R Chart

[‡]For attribute data: c Chart

[††]For attribute data: p Chart

Note: For a description of other charts refer to a reference on statistical process control (SPC).

– p Chart (attribute data)

– Sample data: Minimum (25) samples, subgroups size may vary (sample size varies). Subgroup size is typically 50 or greater to show defectives per subgroup of 4 or greater.

Note: Subgroup size (n) should be within + or − 20% of the average size or control limits need to be recalculated.

Calculations: See p Chart example.

$$\bar{p} \text{ (Avg.)} = \frac{\text{Total number of defectives}}{\text{Total number of units}} = \frac{183}{1526} = .12 \text{ or } 12\%$$

$$\bar{n} \text{ (Avg.)} = \frac{\text{Total number of units}}{\text{Total number of samples}} = \frac{1526}{25} = 61$$

Upper Control Limit:

$$\text{UCLp} = \bar{p} + 3\sqrt{\frac{(\bar{p} \times (100\% - \bar{p})}{n}}$$

$$= 12 + 3\sqrt{\frac{12 \times (100 - 12)}{61}} = 12 + 3\sqrt{\frac{1056}{61}}$$

$$= 12 + 3\sqrt{17.31} \quad = 12 + 12.48$$

$$= 24.48$$

notes and key points (continued)

Lower Control Limit:

$$\text{LcLp} = \bar{p} - 3\sqrt{\frac{(\bar{p} \times (100\% - \bar{p})}{n}}$$

$$= 12 - 12.48 \text{ (from above)}$$

$$= -.48$$

Note: Often the answer is negative. Therefore the lower control limit is at zero!

step-by-step procedure

STEP 1 Determine the type of attribute control chart to be used. See example *Paint Rejects per Hour* (attribute control chart—type *p*).

STEP 2 Collect at least 25 samples of data; subgroups can vary but must have at least 50 units to show defectives per subgroup of 4 or greater.

STEP 3 Prepare a type *p* chart and continue to record collected data as shown. See example chart.

STEP 4 After all 25 subgroups (samples) have been recorded, perform all required calculations. See *notes and key points* above for example.

STEP 5 Plot and connect plotted points to form a trendline. Verify that the trendline points reflect percentage of defectives.

STEP 6 Finalize and date the chart.

example of tool application

Paint Rejects per Hour

Source: Mfg. **Process:** Painting **Characteristic:** Paint rejects/hour

$\bar{p} = 12$ UCL = 24.5   LCL = 0   **Prepared by:** W.J.M

Frequency: Once/day

Calculation date: xx/xx/xx

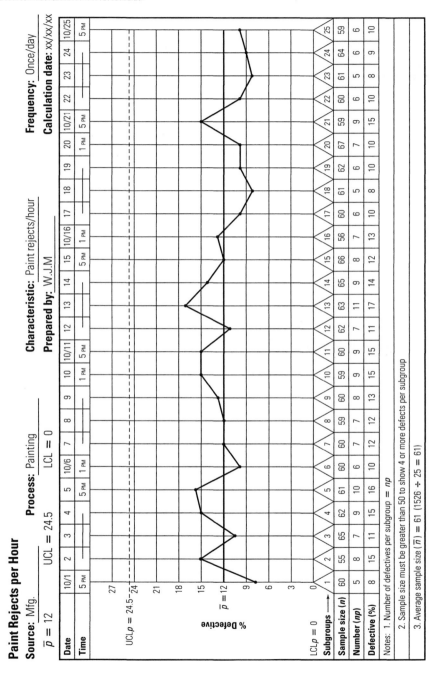

UCLp = 24.5

$\bar{p} = 12$

LCLp = 0

% Defective

Date	10/1				10/6					10/11					10/16					10/21					10/25
Time	5 PM				1 PM					5 PM					1 PM					5 PM					5 PM
Subgroups →	1	2	3	4	5	6	7	8	9	10	11	12	13	14	15	16	17	18	19	20	21	22	23	24	25
Sample size (*n*)	60	55	65	62	61	60	60	59	60	59	60	62	63	65	66	56	60	61	62	67	59	60	61	64	59
Number (*np*)	5	8	7	9	10	6	7	7	8	9	9	7	11	9	8	7	6	5	6	7	9	6	5	6	6
Defective (%)	8	15	11	15	16	10	12	12	13	15	15	11	17	14	12	13	10	8	10	10	15	10	8	9	10

Notes: 1. Number of defectives per subgroup = *np*

2. Sample size must be greater than 50 to show 4 or more defects per subgroup

3. Average sample size ($\bar{n}$) = 61 (1526 ÷ 25 = 61)

Control Chart—$\bar{X}$-R (Variable)

tool description

A control chart is a graph that plots randomly selected data over time in order to determine if a process is performing to requirements or is, therefore, under statistical control. The chart displays whether a problem is caused by an unusual or special cause (correctable error) or is due to chance causes (natural variation) alone.

typical application

- To determine if a process is performing to upper and lower control-limit requirements (process is kept in control).
- To monitor process variations over time, with regard to both special or chance causes.
- To identify opportunities for improving quality and to measure process improvement.
- To serve as a quality measurement technique.

problem-solving phase

➥ Select and define problem or opportunity

➥ Identify and analyze causes or potential change

    Develop and plan possible solutions or change

➥ Implement and evaluate solution or change

➥ Measure and report solution or change results

    Recognize and reward team efforts

links to other tools in *Tool Navigator™*

before

Variance analysis

Sampling methods

Observation

Checksheet

Events log

after

Process capability ratios

Standard deviation

Descriptive statistics

Process analysis

Work flow analysis (WFA)

notes and key points

Types of Control Charts	
Data Required	**For Specific Chart**
Quantitative Variable Data Continuous or measurements *Example:* size, downtime, dimensions, activities per day, etc.	• $\bar{X}$ – R chart[†] (average $\bar{X}$ and range "R" of samples) • $\bar{X}$ – S chart (average $\bar{X}$ and standard deviations "S" of samples)
Qualitative Attribute Data Discrete or counts *Example:* Complaints, rework, missed due dates, delays, rejects, etc.	• c chart[‡] (number of defects in a subgroup) • np chart (number of defective units in a subgroup) • p chart[†‡] (percentage defective) • μ chart (defects per unit)

Most commonly used charts:

  †For variable data: $\bar{X}$-R Chart

  ‡For attribute data: c Chart

  †‡For attribute data: p Chart

Note: For a description of other charts refer to a reference on statistical process control (SPC).

– $\bar{X}$-R Chart (variable data)

– Sample data: Random sampling, minimum (20) samples, minimum (5) data points in each subgroup.

– Calculations: See $\bar{X}$-R Chart example

Table of Factors for $\bar{X}$ & R Charts			
Data Points in Subgroup (*n*)	**Factors for $\bar{X}$ Chart**	**Factors for R Chart**	
	A2	Upper–D3	Lower–D4
2	1.880	0	3.268
3	1.023	0	2.574
4	.729	0	2.282
5	.577	0	2.114
6	.483	0	2.004
7	.419	.076	1.924
8	.373	.136	1.864
9	.337	.184	1.816
10	.308	.223	1.777

notes and key points (continued)

$$\bar{X} = \frac{\Sigma\times}{n}, \bar{X} = \frac{\text{Sum of measurements}}{\text{Number of samples}}$$

$$\bar{R} = H–L, R = \text{Highest} - \text{Lowest measurements}$$

$$\bar{X} = \frac{\Sigma\bar{x}}{k}, \bar{\bar{X}} = \frac{\text{Sum of averages }(\bar{X})}{\text{Number of subgroups}} \quad \bar{\bar{X}} = \frac{99.94}{20}, \bar{\bar{X}} = 51\%$$

$$\bar{R} = \frac{\Sigma R}{k}, \bar{R} = \frac{\text{Sum of ranges }(R)}{\text{Number of subgroups}} \quad \bar{R} = \frac{4.78}{20}, \bar{R} = .24$$

$$UCL_{\bar{x}} = \bar{\bar{X}} + A_2\bar{R}, \ UCL_{\bar{x}} = 5.00 + (.577 \times .24), \ UCL_{\bar{x}} = 5.14$$

$$LCL_{\bar{x}} = X - A_2\bar{R}, \ LCL_{\bar{x}} = 5.00 - (.577 \times .24), \ LCL_{\bar{x}} = 4.86$$

$$UCL_R = D_4\bar{R}, \ UCL_R = 2.114 \times .24, \ UCL_R = .51$$

$$LCL_R = D_3\bar{R}, \ \text{Factor} = 0, \ \text{therefore} \ LCL_R = 0$$

step-by-step procedure

STEP 1 Determine the type of variance control chart to be used. See example *Connector Wire* (variables control chart—type $\bar{X}$-R).

STEP 2 Collect at least 20 samples of data, 5 measurements per sample. Sampling should be random and according to a set frequency over a period of time.

STEP 3 Prepare a type $\bar{X}$-R Chart and record collected data as shown. See example chart.

STEP 4 After all 20 subgroups (samples) have been recorded, perform all required calculations. See *notes and key points* above for example.

STEP 5 Plot and connect plotted points to draw trendlines. Verify that trendline points reflect recorded averages ($\bar{X}$) and ranges (R).

STEP 6 Analyze plotted data for significant variance or patterns.

example of tool application

Connector Wire
Date: xx/xx/xx
Source: E-5

Process: Connector wire

Characteristic: Length: 15 cm

Frequency: Daily

$\bar{\bar{X}}$ = 5.00   UCL = 5.14   LCL = 4.86 $\bar{R}$ = .24   UCL = .51   LCL = 0   USL = 4.8   LSL = 5.2

Time	09:00	10:00	11:00	12:00	14:00	15:00	16:30	17:00	17:30	18:00	18:30	19:00	20:00	21:00	22:30	23:00	23:30	24:00	01:00	02:00
Sample measurements 1	4.81	5.05	5.15	4.95	4.94	4.90	4.91	4.90	5.11	4.99	5.00	4.88	4.99	5.25	4.99	5.01	4.70	5.20	5.04	4.84
2	4.80	5.10	4.83	5.14	5.12	5.13	5.11	5.09	5.12	5.08	5.06	5.14	4.90	5.27	4.95	5.09	4.69	5.18	5.06	4.95
3	4.90	4.99	4.95	5.06	5.04	4.92	5.10	4.82	4.93	4.83	4.90	5.07	4.98	5.18	4.90	4.91	4.70	5.25	5.00	5.11
4	5.11	5.21	4.99	5.01	4.81	5.02	5.17	5.03	5.18	5.04	5.05	5.07	4.98	5.25	5.26	5.00	4.85	5.26	4.97	5.04
5	5.19	4.81	5.08	5.02	4.95	5.03	5.00	5.01	5.02	5.02	4.96	4.84	4.92	5.20	4.70	4.95	4.70	5.30	4.98	4.98
6																				
7																				
Total	24.81	25.16	25.00	25.18	24.86	25.00	24.85	24.85	25.36	24.96	24.97	25.00	24.77	26.16	24.08	24.96	23.64	26.19	25.05	24.92
Avg. ($\bar{X}$)	4.96	5.03	5.00	5.04	4.97	5.00	4.97	4.97	5.07	4.99	4.99	5.00	4.95	5.23	4.82	4.99	4.73	5.24	5.01	4.98
Range (R)	.39	.40	.32	.19	.31	.23	.26	.27	.25	.25	.16	.30	.09	.09	.56	.18	.16	.12	.09	.16

$\bar{X}$ chart:
UCL 5.14
5.20
5.10
5.00
$\bar{\bar{X}}$
4.90
LCL 4.86
4.80

R chart:
UCL .51
56
49
42
35
28
$\bar{R}$ .24
21
14
7
0

Cost of Quality

AKA **Cost of Quality Analysis**

Analyzing/Trending (AT)

tool description

The cost of quality technique can be used as a powerful tool to identify and break down the readily apparent and the more hidden costs of quality activities. Costs are associated with conformance (prevention and assessment activities) and nonconformance (internal and external failures) factors.

typical application

- To identify total quality costs of operations.
- To analyze quality costs for the purpose of discovering cost savings and process improvement opportunities.
- To educate the work force on the cost of quality or lack of quality.

problem-solving phase

Select and define problem or opportunity

➡ Identify and analyze causes or potential change

➡ Develop and plan possible solutions or change

Implement and evaluate solution or change

Measure and report solution or change results

Recognize and reward team efforts

links to other tools in
Tool Navigator™

before

Data collection strategy

Activity cost matrix

Interview technique

Focus group

Starbursting

after

Cost-benefit analysis

Balance sheet

Presentation

Process analysis

Run-it-by

notes and key points

Estimating Cost of Quality (COQ):

$$COQ = \frac{\text{Total time spent on COQ elements}}{\text{Total project time}} \times \frac{\text{Total cost of}}{\text{project time}}$$

step-by-step procedure

STEP 1   A facilitator team brainstorms and lists all quality factors for the purpose of estimating total operational cost.

STEP 2   Team participants receive assignments to collect actual cost data wherever possible. Other costs for quality-related activities need to be estimated by the process owners or department managers.

STEP 3   The facilitator sorts costs into conformance and nonconformance. See example *Cost of Quality (COQ) Measurement*.

STEP 4   The team searches the cost data for the major cost items and discusses alternatives to lower these costs.

STEP 5   A list of potential cost-saving action items is drawn up. This could also be a list of potential process improvements to greatly reduce or eliminate quality-related costs.

STEP 6   Participants use the checklist in this example to generate other ideas for further consideration and analysis.

Cost of Quality (COQ) Measurement

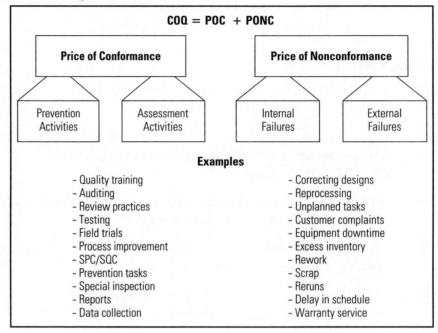

$$COQ = POC + PONC$$

Price of Conformance		Price of Nonconformance	
Prevention Activities	Assessment Activities	Internal Failures	External Failures

Examples

- Quality training	- Correcting designs
- Auditing	- Reprocessing
- Review practices	- Unplanned tasks
- Testing	- Customer complaints
- Field trials	- Equipment downtime
- Process improvement	- Excess inventory
- SPC/SQC	- Rework
- Prevention tasks	- Scrap
- Special inspection	- Reruns
- Reports	- Delay in schedule
- Data collection	- Warranty service

Events Log

AKA **Daily Operations Log**

Changing/Implementing (CI)

tool description

An events log is a historical data trace for critical operations. It provides background information on process variation, changes in operations, supplier problems, and other information useful to problem-solving teams. Entries are made by anyone directly involved in the process and are checked frequently by the process owner.

typical application

- To detect any early trends that may result in a loss of quality or productivity.
- To collect data for the purpose of monitoring operations.
- To maintain a problem prevention procedure.

problem-solving phase

➡ Select and define problem or opportunity

➡ Identify and analyze causes or potential change

➡ Develop and plan possible solutions or change

Implement and evaluate solution or change

Measure and report solution or change results

Recognize and reward team efforts

notes and key points

- Events are considered to be changes in people, equipment, tools, methods, forms, measurements, materials, suppliers, or environment. Frequently occurring problems must be included in the events log for possible future problem-solving activities.

links to other tools in *Tool Navigator*™

before

Observation

Data collection strategy

Checklist

Process analysis

Problem analysis

after

Defect map

Countermeasures matrix

Quality chart

Monthly assessment schedule

Action plan

step-by-step procedure

STEP 1 An events log is started the first day of every month by the unit manager.

STEP 2 Entries are made for every event and each entry is coded for a specific meaning. See example *Work Cell #5—Daily Operations*.

STEP 3 The unit manager checks the events log on a daily basis for variations, patterns, trends, problem areas, etc.

STEP 4 Events logs are used in staff meetings and filed for future reference if needed.

example of tool application

Work Cell #5 — Daily Operations

WC #5 — Product Line: Scope Assembly				Period: 2/1/xx — 3/1/xx
Date	Time	Code	Comments	By
2/1/xx	08:00	2	Absent	J.H.
2/2/xx	09:00	12	Scrap — "D" lens — scratch	A.L.
"	15:30	12	Scrap — "D" lens — 12 lenses scratched	A.L.
2/3/xx	11:30	1	Jay off sick	S.P.
2/4/xx	10:15	14	Inspector found more defective "D" lenses	A.L.
"	13:20	14	Supplier notified by receiving inspection	A.L.
2/5/xx	07:45	11	Delay, parts shortage	A.L.
2/6/xx	15:50	10	Work reassignments (to loan)	J.H.
			includes documents	
2/27/xx	09:30	9	Form 96	R.I.
2/28/xx	14:20	4	Calibration schedule — 2 hrs.	D.O.

Codes – changes		Codes – problems
1 - Operator	7 - Set-up	11 - Material shortage
2 - Expeditor	8 - Process	12 - Scrap
3 - Inspector	9 - Form/drwg.	13 - Rework
4 - Equipment	10 - Other	14 - Defective part
5 - Tooling		15 - Downtime
6 - Material		16 - Control charts

Frequency Distribution (FD)

AKA **Frequency Table**

Analyzing/Trending (AT)

tool description

A frequency distribution can display in table format quantitative (class intervals) as well as qualitative (categories) data organized in a meaningful order. The FD is often used to group data for histogram, pie chart, or other tools.

typical application

- To determine how data is distributed over an acceptable range of upper and lower limits.
- To sort and group raw data.
- To show distribution ratios (percent).

problem-solving phase

➡ Select and define problem or opportunity

➡ Identify and analyze causes or potential change

Develop and plan possible solutions or change

Implement and evaluate solution or change

Measure and report solution or change results

Recognize and reward team efforts

links to other tools in *Tool Navigator™*

before
Checksheet
Observation
Data collection strategy
Surveying
Interview technique

after
Histogram
Linechart
Pie chart
Two-directional bar chart
Trend analysis

notes and key points

Preparation for Grouping of Data

- Determine the range(s) of the distribution

  $R = (H - L) + 1$

- For smaller data sets, $n = \; < 100$: number of class intervals (C.I.) between 5–10

  For larger data sets, $n = \; > 100$: number of class intervals (C.I.) between 10–20.

- Width of the class interval to be 2, 3, 5, 10, 20 for smaller numbers. (Add zeros for larger data sets.)

- Select number of class intervals:

  number of C.I. $= \dfrac{R}{2,\ 3,\ 5,\ 10,\ 20}$ ⟵——— Range

  ⟵——— C.I. Width

- Check if the lowest data point in the data set is divisible an equal number of times as those by the C.I. width. If not, select the next lower data point that is.

step-by-step procedure

STEP 1   Count the number (N) of data points or observations:

16	11	15	22	31	Count = 30	⑩	15	17	21	25	
19	10	24	14	26		11	15	18	22	26	
17	22	16	12	27		12	16	19	22	27	
14	13	16	24	28		13	16	19	22	28	
19	29	21	22	19		14	16	19	24	29	High = 31
18	15	17	19	25		14	17	19	24	㉛	Low = 10

STEP 2   Identify highest and lowest data point.

STEP 3   Calculate range (R): $R = (H - L) + 1$

$= (31 - 10) + 1 = 22$

STEP 4   Determine the number of class intervals (C.I.) and width:

number of C.I. $= \dfrac{22}{2,\ ③,\ 5,\ 10,\ 20} = 7.33$ or 8

width = 3   Must be between 5–10

STEP 5   Construct frequency distribution table to display columns for: class interval, class frequency (f), and relative frequency (rf), which is expressed in percent of total (N).

step-by-step procedure (continued)

STEP 6   Insert prepared data into the FD table. See example *Customer Complaints per Day* (quantitative data). Date the table.

Note: For categorical data, construct a FD table as shown in example for customer response data.

example of tool application

Customer Complaints per Day

A. Customer Complaints/Day		Period = 30 days
Class Interval	Class *f*	Class *rf* (%)
9–11	2	6.67
12–14	4	13.33
15–17	7	23.33
18–20	5	16.67
21–23	4	13.33
24–26	4	13.33
27–29	3	10.00
30–32	1	3.33
Date xx/xx/xx	30	100.00

B. Customer Response Data		Total Respondents = 26
Category	(*N*)	%
Strongly agree	6	23.08
Agree	7	26.92
Neither	3	11.54
Disagree	5	19.23
Strongly disagree	3	11.54
Missing response	2	7.69
Date xx/xx/xx	26	100.00

Line Chart

Analyzing/Trending (AT)

tool description

A simple line chart is an ideal method for showing trends in quality, quantity, cost, customer satisfaction, and so on. It is often a first indication that some problem exists during the monitoring and tracking of quality performance data.

typical application

- To monitor and track data over a period of time.
- To show a trendline analysis.
- To display change in quality performance.
- To identify shifts from predetermined averages.

problem-solving phase

➥ Select and define problem or opportunity

➥ Identify and analyze causes or potential change

    Develop and plan possible solutions or change

➥ Implement and evaluate solution or change

➥ Measure and report solution or change results

    Recognize and reward team efforts

notes and key points

- To enhance the interpretation of a line chart, a "goal for improvement" or a "standard" line should be drawn to verify actual performance to the desired goal or standard.

links to other tools in *Tool Navigator*™

before
Frequency distribution (FD)
Checksheet
Observation
Focus group
Surveying

after
Stratification
Multivariable chart
Trend analysis
Pie chart
Pareto chart

76

40 Top Tools for Manufacturers

step-by-step procedure

STEP 1 Collect data from some source, such as a checksheet. See example *Customer Complaints*.

Type	Week 1					Week 2					Week 3					Week 4					Total
	M	T	W	T	F	M	T	W	T	F	M	T	W	T	F	M	T	W	T	F	
Ordering	||	|	⾞	|	|||	||	||||	||||	||||	⾞	|||	⾞	|	⾞|		⾞	||||	||	⾞	||||	66
Shipping	|	||		||	|	|		|||			||	|		||	|	|	|		||		20
Billing	|	|		|		|		||		|||	|	|			|||	||	|	||		|	20
Defect	||	|	⾞	⾞	||	|	||	||||	|	|	||		||	⾞		|	|	||	|	|||	41
Service	⾞	|	||||	|||	⾞	||	|	⾞	|	⾞	⾞	⾞	||	⾞	|	⾞|	||	⾞	|	⾞	69
Total	11	6	14	12	11	7	7	18	6	14	13	12	5	18	5	15	9	11	9	13	216

STEP 2 Construct a line chart and apply the 3:4 ratio rule: The height of the Y axis must be 75 percent of the length of the X axis.

STEP 3 Label the axes, plot and connect the data points, and draw the line as encoded in the legend. Date the line chart.

example of tool application

Customer Complaints

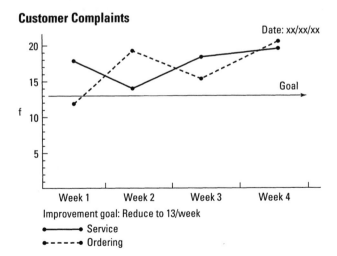

Date: xx/xx/xx

Goal

Improvement goal: Reduce to 13/week

●——● Service
●- - - -● Ordering

Multivariable Chart

AKA **Multi-Var Chart**

Analyzing/Trending (AT)

tool description

A multivariable chart is used to measure time-series data of multiple variables reflecting process capability variance. This chart provides process variable correlation and interaction information that is not usually found when examining traditional control charts one at a time.

typical application

- To construct an overlay of certain process variables normally recorded on control charts.
- To allow time-series analysis of process variables.
- To identify possible problem causes.
- To contribute to design of experiments (DOE) and statistical process control (SPC) activities.

problem-solving phase

➡ Select and define problem or opportunity

➡ Identify and analyze causes or potential change

Develop and plan possible solutions or change

Implement and evaluate solution or change

Measure and report solution or change results

Recognize and reward team efforts

notes and key points

- Note that it is difficult plotting process variables along matching time spans. Also, scaling of upper and lower specification limits (USL-LSL) for process variables may be limited to the base variable with the greatest upper and lower deviation from the specification target value.

**links to
other tools in
Tool Navigator™**

before

Control chart

Data collection strategy

Checksheet

Checklist

Standard deviation

after

Variance analysis

Process capability ratios

Analysis of variance

Potential problem analysis (PPA)

Trend analysis

step-by-step procedure

STEP 1 First, acquire the target and upper and lower specification values from design engineering, manufacturing, or the quality department.

STEP 2 Identify two to four related process variables. See example *Painting Quality*.

STEP 3 Draw a chart, with the center line labeled *spec* (target) value and upper and lower horizontal lines designated *USL* and *LSL* respectively.

STEP 4 Designate the *x*-axis with the proper time scale. The *x*-axis represents an amount of time for the variable with the longest time span.

STEP 5 Identify process variables and encode for plotting and analysis purposes.

STEP 6 Take measurements and plot by connecting data points.

STEP 7 Date the chart and keep for later reference.

example of tool application

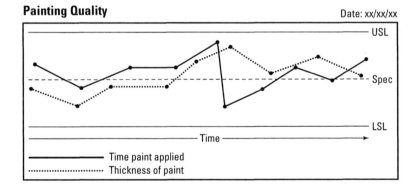

Process Capability Ratios

AKA **Capability Indices**

Analyzing/Trending (AT)

tool description

Process capability ratios are calculated to determine the process variation that comes from natural or special causes. These ratios, also called C_p and C_{pk} Indices, relate the process variability to the design specification (tolerance) that reflects the customer's expectation or requirements. C_p is listed to characterize the capability, and C_{pk} is used to measure actual process performance.

typical application

- To estimate how well the process meets customer requirements.
- To monitor and measure product quality.
- To verify process variability to design specifications.
- To promote communications among design engineering, suppliers, and manufacturing.

problem-solving phase

➡ Select and define problem or opportunity

➡ Identify and analyze causes or potential change

  Develop and plan possible solutions or change

  Implement and evaluate solution or change

➡ Measure and report solution or change results

  Recognize and reward team efforts

links to other tools in *Tool Navigator™*

before

Standard deviation

Sampling methods

Descriptive statistics

Normal probability distribution

Control chart

after

Variance analysis

process flowchart

Process analysis

Activity analysis

Basili data collection method

notes and key points

- Definitions of process capability ratios:
  - C_p: A measure of ideal or potential process capability. The C_p index reflects the best ability of a process to perform within lower and upper design specification limits (LSL $\leftrightarrow$ USL).
  - C_{pk}: A measure of actual or located process performance. The C_{pk} index reflects the actual, located process mean relative to the design target value.
  - CPU: Upper process capability
  - CPL: Lower process capability
  - μ or $\overline{\overline{X}}$: Process average

notes and key points (continued)

- Equations for calculating C_p and C_{pk}:

  (A) $C_p = \dfrac{\text{Design tolerance}}{\text{Process variation}}$, $C_p = \dfrac{\text{USL} - \text{LSL}}{\pm 3\sigma}$, $C_p = \dfrac{\text{USL} - \text{LSL}}{6\sigma}$

  or

  (B) $\text{CPU} = \dfrac{\text{USL} - \mu}{3\sigma}$, $\text{CPL} = \dfrac{\mu - \text{LSL}}{3\sigma}$

  (A) $C_{pk} = C_p (1 - k)$ Where $k = \dfrac{\left| \begin{array}{ccc} \text{Target} & - & \text{Actual} \\ \text{mean point} & & \text{mean point} \end{array} \right|}{1/2 \,(\text{USL} - \text{LSL})}$

  or

  (B) $C_{pk} = \text{Minimum of (CPU, CPL)}$

- Process capability:

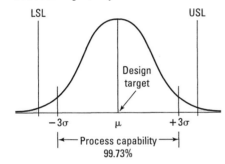

USL = Upper specification limit
LSL = Lower specification limit
μ = Design target mean (often shown
           as process averge $\bar{\bar{X}}$)
$\pm 3\sigma$ = 99.73% of area under the normal
           probability distribution curve

step-by-step procedure

STEP 1   A team is formed with the goal of reduced process variability.

STEP 2   Through the data collection process, process data such as specific product specifications tolerances, control charts, samples of measurements, and historical data calculations, are acquired. See example *Fax Paper Extension Wire Measurements*.

STEP 3   Sample data are organized and summarized using descriptive statistics. A recommendation is made to read some background tools contained in this book:

- Descriptive statistics (tool 66)
- Normal probability distribution (tool 119)
- Standard deviation (tool 184)

step-by-step procedure (continued)

STEP 4   Using process means and standard deviations, the C_p capability ratios can be calculated and compared.
Note: Measurement in this example is centimeters (cm).

STEP 5   An action plan is developed on the basis of the actual process performance results calculated. Required activities usually require reduction of process variations or recentering of the process within the specifications.

example of tool application

Fax Paper Extension Wire Measurements

C_p — Process capability and variability

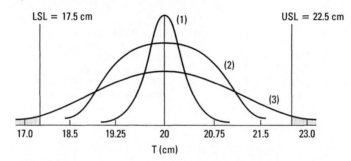

– Distribution #1: This is an *ideal* or *best* process.

$$C_p = \frac{22.5 - 17.5}{1.5}$$

$\sigma = .25, \pm 3\sigma = 1.5$

$$C_p = 3.33$$

– Distribution #2: This is a *capable* process.

$$C_p = \frac{22.5 - 17.5}{3}$$

$\sigma = .50, \pm 3\sigma = 3$

$$C_p = 1.67$$

– Distribution #3: This is an *uncapable* process.

$$C_p = \frac{22.5 - 17.5}{6}$$

$\sigma = 1, \pm 3\sigma = 6$

$$C_p = .83$$

Note: Shaded areas outside the specification limits display nonconformance to design specs.

example of tool application   (continued)

C_{pk} — Actual or Located Process Performance

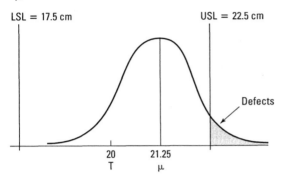

LSL = 17.5 cm   USL = 22.5 cm

Defects

20   21.25
T   μ

$$C_{pk} = C_p (1 - K) \qquad K = \frac{|\; 20 - 21.25 \;|}{1/2 \,(22.5 - 17.5)}$$

$$C_{pk} = 1.67 \,(1 - .5) \qquad K = \frac{-1.25}{2.50}$$

$$C_{pk} = .83 \qquad K = .5$$

Alternate method of calculating C_p, C_{pk}:

– Distribution #2: *Capable* process.

$$CPU = \frac{22.5 - 20}{1.5} \qquad CPL = \frac{20 - 17.5}{1.5}$$

$$CPU = 1.67 \qquad CPL = 1.67$$

– Distribution #3: *Uncapable* process.

$$CPU = \frac{22.5 - 21.25}{1.5} \qquad CPL = \frac{21.25 - 17.5}{1.5}$$

$$CPU = .83 \qquad CPL = 2.5$$

– Check: $\dfrac{.83 + 2.5}{2} = 1.67$

C_{pk} = Minimum of (83, 2.5)
C_{pk} = .83 as calculated above.

Stratum Chart

AKA **Surface Chart, Sarape Chart**

Analyzing/Trending (AT)

tool description

A stratum chart is an effective means to demonstrate cumulative additions of data that range from low to high and that are plotted along a horizontal time scale. Coloring or shading is used to differentiate among variables and to provide a quick interpretation of accumulation and relationships between plotted data.

typical application

- To show cumulative changes in data over time.
- To display the effect of plotted variables, each variables' gradual change over time and relationships to each other.

problem-solving phase

➡ Select and define problem or opportunity

➡ Identify and analyze causes or potential change

Develop and plan possible solutions or change

➡ Implement and evaluate solution or change

Measure and report solution or change results

Recognize and reward team efforts

notes and key points

- Line of curves cannot overlap.
- Coloring or shading is needed to demonstrate the effect of cumulative changes in data plotted.

**links to
other tools in
Tool Navigator™**

before

Data collection strategy

Checksheet

Frequency distribution (FD)

Trend analysis

Sampling methods

after

Major program status

Process analysis

Monthly assessment schedule

Information needs analysis

Presentation

step-by-step procedure

STEP 1   Collect historical data for variables to be plotted. See example *TQM-Related Start-up Costs*.

STEP 2   Scale the stratum chart to allow for highest cumulative number.

STEP 3   Plot the data along the horizontal time scale as shown in the example.

STEP 4   Color or provide pattern shading for variable data plotted.

STEP 5   Check the stratum chart for accuracy and provide date of issue.

example of tool application

TQM Related Start-up Costs

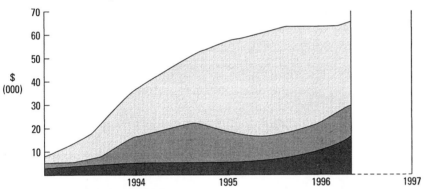

Note: Scale in thousands of dollars
- TQM teams – cost of meetings
- TQM tools training costs
- Facilities/supplies costs

Variance Analysis

AKA **Variance Matrix**

Analyzing/Trending (AT)

tool description

A variance analysis tool discovers key process variances or deviations from specifications in a selected process for the purpose of reducing or controlling the serious impact these variances have on service or product quality. In addition, variance analysis allows a closer look at process cycle time, waste, rework, supplier quality, and overall costs.

typical application

- To utilize a variance analysis tool after a team has completed process mapping in problem-solving efforts.
- To document key variances that have an overall negative affect on product quality.
- To identify process capability and deviations from customer requirements.
- To perform continuous process improvement and cycle time reduction activities.

problem-solving phase

➡ Select and define problem or opportunity
➡ Identify and analyze causes or potential change
  Develop and plan possible solutions or change
  Implement and evaluate solution or change
➡ Measure and report solution or change results
  Recognize and reward team efforts

links to other tools in *Tool Navigator™*

before

Brainstorming

Sandard deviation

Process capability ratios

Control chart

Process mapping

after

Pareto diagram

Cause and effect diagram (CED)

Analysis of variance

Activity analysis

Shewhart PDCA cycle

notes and key points

- Variance analysis matrix construction:
  - Ensure that all process steps are included and sequenced properly in the matrix. A process map will provide this information.
  - Use a consistent level of detail in listing all variances.
  - Focus on identified key variances that potentially can cause the greatest impact (problems) on the process.
  - Variance impact scale: 5 = *high*, 3 = *medium*, 1 = *low*
  - $\boxed{12}$ = variance #12
  - $\textcircled{13}$ = key variance
  - Key variance control table.

Key Variance Control Table Examples

Key Variance	Root Cause?	Where Observed?	Where Controlled?	How Changed
#3–Old issue parts list	Incorrect documents	Supply/kitting	Engineering repro	Revise documents
#10–Panel wood quality	Defective wood	Parts assembly	Customer complaints	Change supplier
#21–Loose connections	Untightened nuts	Final testing	Assembly operator	Operator training

step-by-step procedure

STEP 1 As a first step, the team process maps a selected process so that all participants acquire a greater understanding of the process.

STEP 2 Next, the team reaches consensus on the input and output requirements based on design criteria and customer expectations.

STEP 3 The team's facilitator explains the concepts of a variance analysis and draws a variance analysis matrix on a whiteboard or on flip charts.

STEP 4 Participants agree on the major process steps and sequentially list all known or potential process variances.

STEP 5 The team establishes variance impact criteria and assigns numerical values to variances that would affect other activities. See example *Assembly of Control Panel Simulators*.

STEP 6 Key variances are identified by adding all impact values underneath each listed variance. The highest totals are considered key variances and their respective variance numbers are circled, as shown in this example.

STEP 7 Next, a variance control table is constructed. This table is used to hold all team-identified key variances.

STEP 8 Systematically, the team completes this table. An example is shown in *notes and key points*. The completion of this table will greatly enhance a team's ability to reduce or at least control the key variances that impact the quality of products or services.

STEP 9 Finally, the variance analysis matrix is dated and presented to the process owners.

example of tool application

Assembly of Control Panel Simulators

Process Steps																										Variances	
	1	2	3	4	5	6	7	8	9	10	11	12	13	14	15	16	17	18	19	20	21	22	23	24	25		Matrix date: xx/xx/xx
Instructions	(1)																										Old issue instructions
	5	2																									Unclear instructions
Parts Check	5	1	(3)																								Old issue parts lists (previous model)
			5	4																							Wrong toggle switches (2-way vs. 3-way)
			5		(5)																						Wrong type of lampcaps
			5			6																					Wrong color of lampcaps
			5				7																				Missing fuse (AA)
			1					8																			Missing washers
			3						9																		Wrong size nuts
Parts Assembly										(10)																	Panel wood quality (knots)
											11																Panel sizing (holes don't line up)
										5	1	12															Panel finish (rough edges)
										3	5	3	13														Uneven panel cutting
											5	3	1	14													Uneven stain
										5		5			15												Lampholes too large (diameter)
																16											Bent connectors
	3																17										Wires not tagged
Final Inspection	3	1																18									Missing final inspection tags
												1	3	3					19								Scratched panel finish
																	3			20							Selector switch hard to rotate
Final Testing												3	3								(21)						Loose connections
																					5	(22)					Open varistor
				5																			23				Shorted switch terminals
																					5	5		24			Open circuit
																					5	5	5	5	25		Intermittent operation
Key Variances	16	1	25	5	10	—	—	5	3	16	7	9	4	—	—	—	8	—	—	—	15	10	5	5	5		

Note: Impact scale–5 = high, 3 = medium, 1 = low

"Old issue parts list (previous model)" the key variance with the highest count (25), could cause the greatest impact.

Navigator Tools for Problem Solving

Cause and Effect Diagram Adding Cards (CEDAC)

AKA **Bulletin Board Fishbone**

(**Analyzing/Trending (AT)**)

tool description

The cause and effect diagram adding cards (CEDAC) is a fishbone diagram that typically displays major, generic categories such as people, methods, materials, equipment, measurement, and environment that cause an effect, often perceived as a problem. This diagram is used to systematically analyze cause and effect relationships and to identify potential root causes of a problem. An additional feature of adding cards by those outside the team allows the capture of more ideas from others in an expanded involvement in the problem-solving process. Once the basic diagram is completed and posted, cards or notes indicating more causes or ideas are attached.

typical application

- To assist a team in reaching a common understanding of a complex problem and to share this information with others for more input.
- To expand the team's thinking and to consider all potential causes.
- To post and share a completed cause and effect diagram (CED) for the purpose of allowing others to add potential causes or ideas.
- To define the major categories or sources of root causes.
- To organize and analyze relationships and interactive factors.
- To identify factors that could improve a process.

problem-solving phase

➡ Select and define problem or opportunity

➡ Identify and analyze causes or potential change

Develop and plan possible solutions or change

Implement and evaluate solution or change

Measure and report solution or change results

Recognize and reward team efforts

links to other tools in *Tool Navigator™*

before

Cause and effect diagram (CED)

Brainstorming

Brainwriting pool

6-3-5 method

Five ways

after

Problem specification

Work flow analysis (WFA)

Process analysis

Countermeasures matrix

Pareto chart

notes and key points

- This tool is an expansion of the cause and effect diagram (CED).
- Generic category designations may be substituted. Example: *Procedures* for *methods*, or *facilities* for *requirement*, etc.
- Do not overload categories. Establish another category if more detail is desired.

step-by-step procedure

STEP 1 Reach consensus on a problem to be analyzed. See example *Missed Reproduction Schedules*.

STEP 2 Determine the major categories and place one in each category box.

STEP 3 Brainstorm possible causes for each category and enter in a fishbone fashion by drawing arrows to the main arrow (category), as shown in the example.

STEP 4 Continue to ask questions using the Five Whys tool to search for root causes. Insert and connect potential causes to the various other contributing factors.

STEP 5 When the team feels that the diagram is complete, a final and much larger diagram is drawn.

STEP 6 The diagram is posted in a hallway or on bulletin boards with an invitation for others to examine and possibily add their causes or ideas on available cards or Post-its to the respective categories on the diagram.

STEP 7 After a specified period of time, the diagram is removed and revised to include the additional information. A completed, smaller diagram is reposted with a thank you note.

STEP 8 The team now advances to the next step of further analysis, additional data collection, and problem solving.

example of tool application

Missed Reproduction Schedules

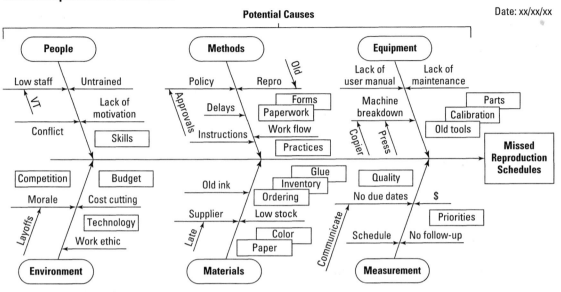

Potential Causes

Date: xx/xx/xx

Checksheet

AKA **Tally Sheet**

(**Data Collecting** (DC))

tool description

A checksheet is a simple form designed to record and quantify facts and data over a period of time. The construction of the checksheet should be tailored to collect data on specific categories and location of defects, frequency of events, or possible causes.

typical application

- To observe an operation and record specific data over a period of time.
- To acquire a short-term observation of process variability on the current situation.
- To identify what potential problem should be addressed first.
- To confirm the effects of problems.

problem-solving phase

➡ Select and define problem or opportunity

➡ Identify and analyze causes or potential change

Develop and plan possible solutions or change

➡ Implement and evaluate solution or change

Measure and report solution or change results

Recognize and reward team efforts

links to other tools in *Tool Navigator*™

before

Data collection strategy

Observation

Surveying

Interview technique

Focus group

after

Stratum chart

Histogram

Frequency distribution (FD)

Box plot

Pareto chart

notes and key points

Types of frequently used checksheets:
- To count occurrences (tally 卌).
- To measure activities (amounts, time, etc.)
- To locate problems or defects (defect map)

step-by-step procedure

STEP 1   Identify data to be collected. See example *Customer Complaints*.

STEP 2   Design checksheet for easy data capture.

STEP 3   Collect data for the stated time period. Example: 20 days.

STEP 4   After a specified period, total the check marks and input this data into the problem-solving process. Date the checksheet.

STEP 5   Note: For another type of checksheet, see the defect map approach.

example of tool application

Customer Complaints   Date: xx/xx/xx

Type	Week 1					Week 2					Week 3					Week 4					Total
	M	T	W	T	F	M	T	W	T	F	M	T	W	T	F	M	T	W	T	F	
Ordering	II	I	卌	I	III	II	IIII	IIII	IIII	卌	III	卌	I	卌I		卌	IIII	II	卌	IIII	66
Shipping	I	II		II	I	I		III			II	I		II	I	I	I		II		20
Billing	I	I		I		I		II		III	I	I			III	II	I	II		I	20
Defect	II	I	卌	卌	II	I	II	IIII	I	I	II		II	卌		I	I	II	I	III	41
Service	卌	I	IIII	III	卌	II	I	卌	I	卌	卌	卌	II	卌	I	卌I	II	卌	I	卌	69
Total	11	6	14	12	11	7	7	18	6	14	13	12	5	18	5	15	9	11	9	13	216

Defect Map

tool description

A defect map displays the location of defects and simplifies the process of data collection and repair. Problem locations are marked so that repair staff know where to look.

typical application

- To point to the location of defects or problems on rejected products.
- To mark or check off the locations of defects on an assembly diagram for the purpose of collecting frequency data of the various defects observed.

problem-solving phase

➡ Select and define problem or opportunity

➡ Identify and analyze causes or potential change

Develop and plan possible solutions or change

Implement and evaluate solution or change

Measure and report solution or change results

Recognize and reward team efforts

links to other tools in *Tool Navigator*™

before

Data collection strategy

Checksheet

Checklist

Observation

Quality chart

after

Pareto chart

Problem specification

Failure mode effect analysis

Potential problem analysis (PPA)

Countermeasures matrix

notes and key points

- Use checksheets to record and summarize various defect map data.

step-by-step procedure

STEP 1 A defect map can be drawn for specific parts, assemblies, or complete units of product. Assembly drawings can also be used as defect maps. See example *Sub-Assembly No. 314—Location of Defects*.

STEP 2 During inspection of parts, assemblies, or units, the observed location of defects is marked on defect maps.

STEP 3 Defect types, locations, and frequencies are summarized on a checksheet as shown in the example.

STEP 4 Use this type of historical defect data to prepare a problem specification statement.

STEP 5 Provide notes on drawings, and date all documentation.

example of tool application

Sub-Assembly No. 314 – Location of Defects

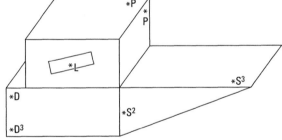

SA – 314	Date: xx/xx/xx
Defects	**Total**
D = Dent S = Scratch P = Paint L = Label	4 5 2 1
Total	12

Note: D^3 stands for three dents in specified area.

Dendrogram

AKA **Reviewed Dendrogram**

Evaluating/Selecting (ES)

tool description

The dendrogram displays, in a tree-type classification format, clusters of characteristics or ideas to be analyzed for potential breakthroughs in product design and development. It can also be used to detail possible solutions to problems or examine process improvement opportunities.

typical application

- To search for potential product innovations.
- To break down and classify large data sets.
- To review and question ideas for problem resolution or process improvement.

problem-solving phase

➡ Select and define problem or opportunity

➡ Identify and analyze causes or potential change

➡ Develop and plan possible solutions or change

Implement and evaluate solution or change

Measure and report solution or change results

Recognize and reward team efforts

links to other tools in *Tool Navigator*™

before

Cluster analysis

Brainwriting pool

Crawford slip method

Delphi method

Focus group

after

House of quality

Activity analysis

Factor analysis

Opportunity analysis

Creativity assessment

notes and key points

- Dendrograms are often used to display the outcomes or results of cluster analyses.

step-by-step procedure

STEP 1   The team facilitator describes the use of a dendrogram and asks the team to brainstorm items within an area of interest. See example *Development of a Better Classroom Pointer.*

STEP 2   The facilitator draws the dendrogram on a whiteboard as the participants further break down a selected characteristic or idea.

STEP 3   The participants discuss preferred ideas and select one for product innovation or problem analysis, as shown in this example.

STEP 4   The participants review the flowdown of characteristics or ideas and date the dendrogram.

example of tool application

Development of a Better Classroom Pointer

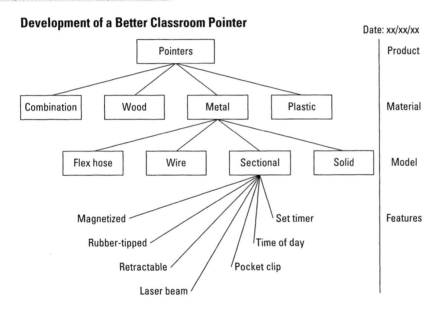

Date: xx/xx/xx

Failure Mode Effect Analysis (FMEA)

AKA **N/A**

Changing/Implementing (CI)

tool description

A failure mode effect analysis (FMEA) is a technique that allows a cross-functional team to identify potential failure modes or causes of failures that may occur as a result of design or process deficiencies. This analysis, furthermore, produces estimates of the effects and level of severity of failures, and it provides recommendations for corrective design of process changes.

typical application

- To consider potential failure modes, causes, effects, and corrective action to be taken.
- To predict the reliability of complex products and processes.
- To assess the impact of failures on internal and external customers.
- To identify ways for a product or subsystem to fail meeting specifications.

problem-solving phase

➡ Select and define problem or opportunity

➡ Identify and analyze causes or potential change

➡ Develop and plan possible solutions or change

Implement and evaluate solution or change

Measure and report solution or change results

Recognize and reward team efforts

links to other tools in Tool Navigator™

before

Fault tree analysis (FTA)

Process flowchart

Process analysis

Activity analysis

Variance analysis

after

Countermeasures matrix

Defect map

Dendrogram

Pareto chart

Events log

notes and key points

Failure mode effect analysis (FMEA) applications:

- *Design FMEA:* Covers all new designs and major design changes to existing products or systems. Considerations: Process capability, assembling space/access for tooling, performance (test) design deviations.
- *Process FMEA:* Covers all planned manufacturing of assembly processes. Considerations: Manufacturing/fabrication, assembly, receiving/inspection, testing/inspection.

Optional scales for FMEA applications:

- Probability of failure *occurrence* (1–10): 1 = remote chance of failing, 10 = very high chance of failing.
- Degree of failure *severity* (1–10): 1 = not noticeable to the customer, 10 = critical failure, probable loss of customer.
- Probability of failure *detection* (1–10): 1 = extremely low chance of escaping defects; 10 = very high chance of escaping defects.
- *Risk Assessment:* The higher the risk priority number (RPN), the more important is the task to eliminate the cause of failure.

RPN = *occurrence* rating × *severity* rating × *detection* rating.

step-by-step procedure

STEP 1   A cross-functional team determines the potential failure modes of a design or process. See example *Power Relay Switch Design and Assembly.*

STEP 2   FMEA forms are distributed to the team's participants and all items on the form are explained and discussed.

STEP 3   Scales and ratings for failure occurrence, severity, and detection are agreed upon and respective rating values assigned for each failure mode, as shown in the example.

STEP 4   The FMEA matrix is completed and checked for accuracy. Finally, recommended action items are assigned.

STEP 5   The finalized FMEA matrix is dated and presented to the process owner.

example of tool application

Power Relay Switch Design and Assembly

Design/Process: Power Relay Assembly								Date: xx/xx/xx	Issue No. 1	
Component and No.: Relay Coil								Team: Power Operators		
Department: Manufacturing								Process Owner: J.K. Nelson		
Potential mode of failure	Potential cause of failure	Potential effect of failure	Design/ process controls	Probablitity of failure occurance	Degree of failure severity	Probablitity of failure detection	Risk priority number	Recommended preventive action	Results or comments
Open winding	Broken coil wire	Relay does not operate	Cp, SPC	2	10	1	20	Pretest	Approved
Intermittent shorting	Lack of insulation	Burn out relay	Visual inspection	3	6	4	72	Change insulation	Assigned to J.P.

*Note: Scales are 1–10 for occurence, severity, and detection ratings.
See description of rating scales for additional information.*

Fault Tree Analysis (FTA)

Analyzing/Trending (AT)

tool description

The fault tree analysis (FTA) was first introduced by Bell Laboratories. It is a logical tool that assists in the uncovering of potential root causes of defects or equipment failures. It can, however, be applied to administrative areas as well. Similar to a tree diagram, the output is the result of various levels of contributing factors or potential causes for failure.

The logic behind this diagram uses *and* and *or* function gates to illustrate symptoms of failure down to real root causes. As shown in the fault tree analysis, an *or* (+) function gate will, for example, have an output of *few inquiries* as a result of input *low interest, lack of info,* or both. On the other hand, an *and* (•) function gate will show a true output of *unprepared instructor* only if all *and* inputs are true. This type of analysis, therefore, provides great insight into the interrelationships among the various cause and effect conditions.

typical application

- To allow a backward approach to systematically identify potential causes of failures.
- To provide an overview of interrelationships between causes and failures.
- To break down failure indications into more detailed input branches.

problem-solving phase

➡ Select and define problem or opportunity

➡ Identify and analyze causes or potential change

➡ Develop and plan possible solutions or change

Implement and evaluate solution or change

Measure and report solution or change results

Recognize and reward team efforts

links to other tools in *Tool Navigator™*

before

Problem specification

Checklist

Failure mode effect analysis

Dendrogram

Events log

after

Countermeasures matrix

Potential problem analysis (PPA)

What-if analysis

Process analysis

Problem analysis

notes and key points

The fault tree is drawn using two logic gates interconnected in various functions.

Three input *and* gate, which requires all three inputs present to have an output.

Two input *or* gate, which requires at least one input to be present to have an output.

Note: This tool works well with other major tools, such as quality function deployment, metrics, and management by policy.

step-by-step procedure

STEP 1   The first step for a team is to write down the failure as an output for the top *or* gate (Level 1). See example *Problem! An Ineffective Training Program for Integrated Product Development Teams (IPDT)*.

STEP 2   Determine what input (symptoms) could be considered a contributing element (Level 2).

STEP 3   Continue breaking down the failure with additional gate levels (Levels 3, 4 or more).

STEP 4   Always ask yourself: Can this gate output be true with any input (*or* function), or does all input need to be present for a true gate output (*and* function)?

STEP 5   Lastly, finalize and date the diagram.

example of tool application

Problem: An Ineffective Training Program for Integrated Product Development Teams (IPDT)

Date: xx/xx/xx

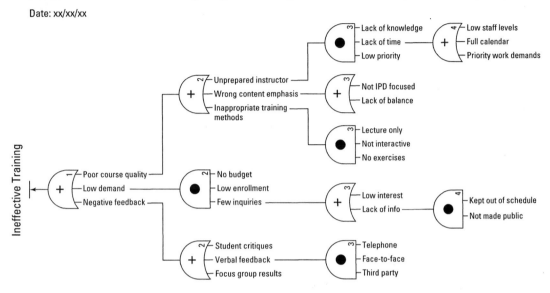

Matrix Diagram

34

(**Planning/Presenting** (PP))

tool description

A matrix diagram is a planning tool that displays two or more sets of characteristics, functions, ideas, or issues. The scanning and comparing of items results in relationship "connections" or cause and effect interactions that can be useful in problem, opportunity, or task requirement analysis.

typical application

- To assign responsibility for action ideas.
- To identify opportunities for improvement.
- To search for possible problem causes.
- To compare the respective strengths of alternative choices.
- To match functions with resource needs.

problem-solving phase

Select and define problem or opportunity
➡ Identify and analyze causes or potential change
Develop and plan possible solutions or change
➡ Implement and evaluate solution or change
Measure and report solution or change results
Recognize and reward team efforts

links to other tools in *Tool Navigator*™

before

Affinity diagram

Tree diagram

Interrelationship digraph (I.D.)

Work Breakdown Structure (WBS)

Process flowchart

after

Matrix data analysis

Activity network diagram

Process decision program chart

Attribute listing

Customer acquisition-defection matrix

notes and key points

"L" shaped   "T" shaped   "Y" shaped   "X" shaped   "C" shaped

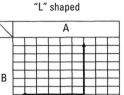 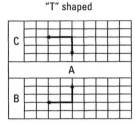 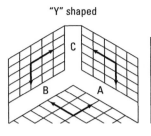 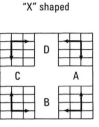 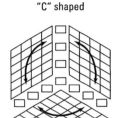

Matrix format	Sets of items	Application/notes	Sets
"L"	2	Check for related items Most frequently used matrix	A → B
"T"	3	Check for related items Two combined "L" matrices	A < B C
"Y"	3	Check for interactions among (3) sets of items	A → B, C
"X"	4	Check for interactions among (4) sets of items	A → B ↑ ↑ D ← C
"C" (cube)	3	Check for linkage between A-B-C	A B C

Optional symbols — some examples

⊙ = High = 9
○ = Medium = 3
△ = Low = 1

⊙ = Primary responsibility
○ = Secondary responsibility
△ = Information only

● = Strong relationship
○ = Medium relationship
X = Weak relationship

step-by-step procedure

STEP 1   Collect two or more sets of items from a brainstorming list, tree diagram, affinity diagram, or other source.

STEP 2   Select a particular matrix format. See example *Improvement Tools Application by Function*.

STEP 3   Construct a matrix diagram and insert sets or items.

STEP 4   Select a set of symbols to show relationship or connection.

STEP 5   Identify relationships, agree on the strength (use the appropriate symbol) of the relationship, and place the symbol at the intersecting points on the matrix.

STEP 6   Verify that all items have been changed and date the matrix diagram.

example of tool application

Improvement Tools Application by Function

Date: xx/xx/xx / Tools / Function	QFD	SPC	JIT	CTM	BPR	CSA	Survey	Benchmarking	DOE	IPD	Metrics
Research	○	○				○	◉	◉	◉	△	
Planning	◉			○	◉			◉		◉	△
Finance											◉
Project mgmt.		△	△	◉			△	○		◉	◉
Administration				◉							◉
Engineering	○	○	△	△	◉			○	◉	○	○
Manufacturing		◉	◉	◉	△			△	○	△	○
Quality	◉	◉	○	△		◉	△	○	△	△	◉
Marketing	○		△			◉	◉	○		△	○
Sales/service	○	△	△	△		◉	◉	△		△	○

◉ = High   ○ = Medium   △ = Low

Note: QFD = Quality function deployment
       SPC = Statistical process control
       JIT = Just-in-time
       CTM = Cycle time management
       CSA = Customer service analysis
       DOE = Design of experiments
       IPD = Integrated product development

Pareto Chart

AKA **Pareto Analysis, Pareto Principle**

Analyzing/Trending (AT)

tool description

The Pareto chart is a bar chart arranged in a descending order of size or importance from left to right to separate and display the critical few from the trivial many causes of a problem. It is named after Vilfredo Pareto who, in the late 1800s, postulated the 80/20 role, which states that 80 percent of the trouble is due to 20 percent of the causes. The Pareto chart will also show the cumulative percentage for each cause on the chart.

typical application

- To prioritize potential causes of a problem.
- To establish and verify cause and effect.
- To reach consensus on what needs to be addressed first.
- To identify improvement opportunities.
- To measure success of corrective action.

problem-solving phase

➡ Select and define problem or opportunity
➡ Identify and analyze causes or potential change
  Develop and plan possible solutions or change
➡ Implement and evaluate solution or change
➡ Measure and report solution or change results
  Recognize and reward team efforts

**links to
other tools in
Tool Navigator™**

before

Checksheet

Data collection strategy

Sampling methods

Frequency distribution

Yield chart

after

Problem specification

Process analysis

Problem analysis

Process mapping

Action plan

notes and key points

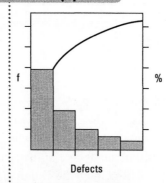

Left vertical scale = frequency of defects
Right vertical scale = percentage of defects
Horizontal scale = types of defects
Trendline = cumulative percentage of defects

step-by-step procedure

STEP 1   Identify the data source. See example *Tape Unit Assembly Defects*.

STEP 2   Calculate percentages for totals of each type of defect.

STEP 3   Draw a Pareto chart as shown in the *notes and key points* above.

STEP 4   Designate left and right vertical scales.

STEP 5   Draw the bars in accordance with total defects per type.

STEP 6   Construct the cumulative line (left to right) by adding the respective percentage for each defect type and connect plots with straight lines, as shown in the example.

STEP 7   Title the Pareto chart; include time period covered and source of data.

example of tool application

Tape Unit Assembly Defects

Tape Unit Assembly Defects (April)	Quantity	%*
A – Unit surface (scars/scratches)	70	47
B – Sub-assembly (defective)	12	8
C – Mounting parts (incorrect/missing)	8	5
D – Insulators (cracked/missing)	23	15
E – Wire (incorrect length)	10	7
F – Assembling (incorrect)	7	5
G – Soldering (% fill rejects)	15	10
H – Labeling (missing/incorrect)	5	3
	150	100

* Rounded

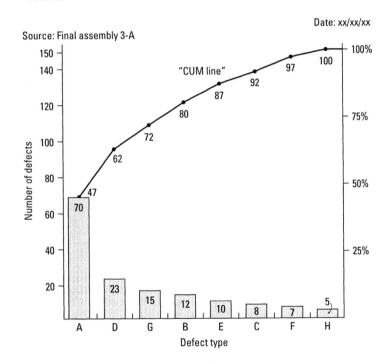

Date: xx/xx/xx

Source: Final assembly 3-A

Potential Problem Analysis (PPA)

AKA **N/A**

Changing/Implementing (CI)

tool description

The potential problem analysis (PPA) is a tool used for minimizing the probability of solution implementation failure by identifying potential problems and possible countermeasures.

typical application

- To anticipate and analyze potential problems.
- To identify potential problems that could cause delays, difficulties, or outright failure in the implementation of a solution.
- To estimate the risk of project failure and the cost of prevention.

problem-solving phase

➡ Select and define problem or opportunity

➡ Identify and analyze causes or potential change

➡ Develop and plan possible solutions or change

Implement and evaluate solution or change

Measure and report solution or change results

Recognize and reward team efforts

links to other tools in *Tool Navigator*™

before

Problem analysis

Activity analysis

Barriers-and-aids analysis

Reverse brainstorming

Action and effect diagram (AED)

after

Action plan

Resource requirements matrix

Consensus decision making

Countermeasures matrix

Facility layout diagram

notes and key points

- The team may identify potential problems as seen from this example.

  What? Parts not available when needed

  Where? At "parts assembly" workstations

  When? First 10 days of each month

  Who? Required by operators of all shifts

  Extend? Could shut down the assembly process

- Suggested definitions and scales.

  R – Risk in implementation (1–10): The amount of risk a potential problem may bring to the implementation of a solution. A rating of 1 stands for very little risk, a rating of 10 may prevent implementation altogether.

  P – Probability of occurrence (1–10): An estimate of probability for a potential problem to materialize. A rating of 1 is very low, 10 is extremely high.

  RP – Residual probability (1–10): A reestimated probability for a potential problem to occur after appropriate countermeasures may prevent a potential problem from appearing. A rating of 1 is very low, 10 is extremely high.

  C – The estimated cost of countermeasures or contingency plans. L = low cost, M = medium cost, H = high cost.

step-by-step procedure

STEP 1 First, a facilitated team determines a set of solution implementation goals and all activities required to successfully reach these goals.

STEP 2 Using the reverse brainstorming method, a list of potential problems is developed and recorded on a whiteboard.

STEP 3 The list of potential problems is used to draw a table of potential problem analysis. See example *Implementation of a Just-in-Time (JIT) Manufacturing System*.

STEP 4 The facilitator asks participants to analyze each potential problem by asking: what? where? when? who? and to what extent? This will provide more detailed information for risk estimation.

STEP 5 Next, the team discusses the amount of risk each potential problem brings to the implementation effort. After reaching consensus on the risk estimates, the probability of problem occurrence is estimated. Ratings are placed with the potential problems in the table, as shown in this example.

step-by-step procedure (continued)

STEP 6 Team discussion follows on the possible prevention of potential problems. Countermeasures are identified, and the residual probability of occurrence estimated. The residual probability is estimated on the basis of a greatly reduced probability of problem occurrence after countermeasures have been placed into effect.

STEP 7 Lastly, contingency plans are agreed upon just in case the countermeasures will not significantly affect the most serious problems.

STEP 8 To complete the potential problem analysis table, costs are estimated for countermeasures and contingency plans.

STEP 9 The team reviews the table, makes final revisions, dates the table, and presents this analysis to the process owners.

example of tool application

Implementation of a Just-in-Time (JIT) Manufacturing System   Date: xx/xx/xx

ID	Possible Causes of Potential Problems	R	P	Countermeasures to Prevent Problems	RP	C	Contingency Plans for High-Risk Problems	C
A	Suppliers cannot meet delivery schedule	8	5	Select and certify more suppliers	2	M	Acquire (purchase) supplier company	H
B	Parts not available when needed	10	7	Rearrange work, balance supply	4	M	Stock critical parts	M
C	Managers show resistance to change	5	4	Get managers involved early	2	L	Rotate, transfer decision-making	L
D	No direct operator involvement	5	9	Establish self-directed work teams	4	M	— — —	
E	Bottlenecks exist; no workflow balance	6	7	Perform process and cycle time analyses	3	H	— — —	
F	Lack of operator JIT training	4	8	Schedule operator JIT training	0	M	Engage outside training firm	M
G	No self-inspection methods in place	3	8	Develop practices and procedures	0	L	— — —	

Notes: R = Risk, P = Probablility, RP = Residual Probability
Scale 1–10, 1 = Low
C = Cost: L = Low, M = Medium, H = High

Problem Specification

AKA **Problem Definition**

(**Planning/Presenting (PP)**)

tool description

The problem specification tool provides team members with a shared understanding of a problem. Moreover it points to an orderly first step of collecting specific, appropriate data for the purpose of writing a problem statement that clearly defines the unacceptable "as is" situation, any process variance, or its potential causes. The problem specification should also describe the "should be" state of the situation or process to be improved.

typical application

- To establish a problem-solving goal or improvement target.
- To clarify a vague condition perceived as a problem.
- To collect data relevant to the problem and possibly indicative of the root causes.
- To satisfy the need for more data.

problem-solving phase

➡ Select and define problem or opportunity
➡ Identify and analyze causes or potential change
　 Develop and plan possible solutions or change
　 Implement and evaluate solution or change
　 Measure and report solution or change results
　 Recognize and reward team efforts

links to other tools in *Tool Navigator™*

before

Data collection strategy

Interview technique

Multivariable chart

Cause and effect diagram (CED)

Pareto chart

after

What-if analysis

Process mapping

Work flow analysis (WFA)

Process analysis

Systems analysis diagram

notes and key points

- A superior problem specification reflects measurable data: Quantitatively expressed data are numbers, percentages, frequencies, time periods, amounts, rate durations, etc. Qualitatively expressed data are perceptions, demographics, or any nominal data scales.

step-by-step procedure

STEP 1   The team starts the problem specifiction process by discussing the current situation; this situation is called the *as is* on the problem specification form. Expand the information to include all recorded data and verbal input. See example *Problem Specification—Quality of Service*.

STEP 2   Next, the preferred situation, called *should be* on the form, is discussed. This ideal state reflects a perceived gap in process performance from the *as is* state.

STEP 3   Close the performance gap between the two states by filling in the information as illustrated by numbers 1 and 2 on the example problem specification form.

STEP 4   Using the information compiled in 1 and 2 on the example, complete the form by providing the appropriate information for the two columns: *problem occurs* 3 and *problem is resolved* 4.

In the final step, develop a final problem statement that encompasses the critical elements of the problem as developed on the form.

STEP 5   The team finalizes the problem statement shown as 5 on the form; team consensus is reached, and the entire team signs off on it.

example of tool application

Quality of Service

Problem Specification—Quality of Service   Date xx/xx/xx

❶ As is situation/condition
Service cycle time is 12 days, the customer
satisfaction index rating is low, and recalls
average 7 per month.

❷ Should be target/goal
Cycle time = 8.5 days, CSI rating = high,
and recalls average 3 per month
(per benchmark data)

When

❸ Problem occurs
– At the end of the month (last six months)
– Missed service calls

❹ Problem is resolved
– Balanced scheduling
– More training
– Concerns for quality

Where

– All service areas
– Business districts

– Timely service regardless of service area

Impact

– 15% increase in customer complaints

– Less than 2% recalls on service calls

People/Groups

– Service department technicians

– Service department technicians
provide better quality service

Related Information

– Pareto analysis and customer satisfaction survey results are available

❺ Final problem statement
The previous six months' service calls schedule produced a 15% increase of customer complaints.
Causes appear to be lengthy cycle time (delays) and quality of service (recalls).

SCAMPER

tool description

The SCAMPER tool is an outcome of the creative facilitation work performed by Alex F. Osborne in the 1950s. Consisting of a checklist of simple questions, this tool can be used by a team to explore the issues and question everything to formulate new, fresh ideas. Problem-solving teams often produce many solution ideas when responding to the SCAMPER questions asked.

typical application

- To question and identify improvement opportunities for processes, products, and services.
- To formulate alternative ideas for problem solving or process change.
- To produce a large number of solution ideas.

problem-solving phase

➡ Select and define problem or opportunity
➡ Identify and analyze causes or potential change
➡ Develop and plan possible solutions or change
　　Implement and evaluate solution or change
　　Measure and report solution or change results
　　Recognize and reward team efforts

links to other tools in *Tool Navigator*™

before
Brainstorming
Checksheet
Defect map
Pareto chart
Events log

after
Starbursting
Countermeasures matrix
Problem analysis
Process analysis
Solution matrix

notes and key points

- The mnemonic SCAMPER (developed by Bob Eberle) stands for:

S - Substitute?

C - Combine?

A - Adapt?

M - Modify? Magnify?

P - Put to other uses?

E - Eliminate? Minimize?

R - Reverse? Rearrange?

step-by-step procedure

STEP 1   Assemble a representative team with knowledge of the topic, issue, or problem to be analyzed. See example *Defective Flashlight Switch*.

STEP 2   One by one, the idea-spurring SCAMPER questions are presented to the team.

STEP 3   Participants discuss the questions and formulate ideas. Responses are recorded as the SCAMPER checklist or questions are completed.

example of tool application

Defective Flashlight Switch

SCAMPER Questions—Defective Switch	Date xx/xx/xx
S – Can we *substitute* a more reliable switch?	
C – *Combine* slide switch assembly with the signaling button?	
A – What ideas or concepts can be *adapted* from other similar switches?	
M – *Modify* the switch to have fewer parts?	
P – Can the switch be *put* to other uses?	
E – Can the switch be *eliminated* or exchanged?	
R – How can we *rearrange* the components of the switch to a more robust design?	

Solution Matrix

39

tool description

The solution matrix is a method to document the various possible solutions for a particular problem. Solutions are rated and ranked to determine the best choice that has the highest chance of success. Criteria for rating is often feasibility, customer value, and change effectiveness.

typical application

- To rate and rank all solutions and identify a workable, best choice.
- To analyze proposed solutions on the basis of significant factors affecting the solution.
- To avoid guessing at the best solution.

problem-solving phase

Select and define problem or opportunity

Identify and analyze causes or potential change

➥ Develop and plan possible solutions or change

Implement and evaluate solution or change

➥ Measure and report solution or change results

Recognize and reward team efforts

notes and key points

- Rating criteria is optional. Care must be taken that rating designations of *high*, *medium*, or *low* will not conflict with other criteria. For example, if solution *implementation time* is *low*, this is a highly (*high*!) desirable characteristic.

links to other tools in *Tool Navigator*™

before

Countermeasures matrix

Decision tree diagram

Consensus decision making

Factor analysis

Different point of view

After

Activity cost matrix

Action plan

Cost-benefit analysis

Balance sheet

Basili data collection method

step-by-step procedure

STEP 1   The team facilitator displays all proposed solutions on a whiteboard or flip chart. A matrix format is used. See example *Reduce the Current Defects per Unit (DPU) Level.*

STEP 2   Participants brainstorm a set of criteria that is recorded into the solution matrix as shown in this example. Consensus is reached on a rating scale.

STEP 3   Each proposed solution is rated against the listed criteria. This process continues until all proposed solutions have been rated.

STEP 4   Next, all rows are totaled by the facilitator and ranked from highest total (rank 1) to lowest total, as shown in this example. The proposed solution ranked 1 is considered the best or preferred solution.

STEP 5   Finally, the completed solution matrix is dated and presented to the process owner.

example of tool application

Reduce the Current Defects Per Unit (DPU) Level

Date xx/xx/xx

Solutions	Change Effectiveness	Implementation Feasibility	Customer Impact	Total	Rank
Train suppliers, ask for C_p data	M	L	H	9	2.5
Implement statistical process control (SPC)	L	H	M	9	2.5
Reduce process cycle time	M	M	L	7	4
Perform C_p studies, benchmark	L	L	L	3	5
Apply robust design principles	H	M	H	13	①

Notes: *high = 5, medium = 3, low = 1*
        Ranking: Highest total is best choice, rank ①

Window Analysis

tool description

The window analysis technique is used to determine the potential root causes(s) of a performance problem. Similar to the Johari window model, this technique questions any two parties, individuals, or organizational units, if a practice, procedure, or a set of work instructions is known and practiced in order to prevent or minimize performance problems.

typical application

- To identify root causes of a performance problem.
- To verify the adherence to company practices, procedures, or work instructions by all organizational units.
- To investigate if performance problems could have been prevented.

problem-solving phase

➡ Select and define problem or opportunity

➡ Identify and analyze causes or potential change

➡ Develop and plan possible solutions or change

Implement and evaluate solution or change

Measure and report solution or change results

Recognize and reward team efforts

links to other tools in *Tool Navigator*™

before

Five whys

Brainstorming

Cause and effect diagram (CED)

Fault tree analysis

Process analysis

after

Problem specification

Countermeasures matrix

Potential problem analysis (PPA)

What-if analysis

Cost-benefit analysis

notes and key points

- Window categories descriptions:
  A: Practices, procedures, or work instructions have been established and both parties (party X and Y) use this information to minimize performance problems. Status: Company directives are followed.
  B: Practices, procedures, or work instructions have been established; however, party X or Y does not use this information correctly. Status: Company directives are not always followed.
  C: Practices, procedures, or work instructions have been established; however, party X or Y does not have this information. Status: Company directives were not communicated to some parties.
  D: Practices, procedures, or work instructions have not been established and, therefore, neither party (party X and Y) has this information. Status: Company directives need to be communicated to make information available to all parties.
- $\overline{\text{Known}}$ = not known (negation of *known*).

step-by-step procedure

STEP 1   A team is assembled with representation from the organizational units to be analyzed. See example *Lack of Adherence to World Class Practices: Quality Function Deployment (QFD), Design of Experiments (DOE), Cycle Time Management (CTM)*.

STEP 2   Participants identify and discuss perceived performance problems. A priority list of no more than five problems is developed.

STEP 3   Using the window diagram, participants systematically explore all squares and discuss each category (A–D) to identify which window square best represents the true situation.

STEP 4   Lastly, participants receive assignments to collect data to prove or disprove a particular category (B–D) previously selected as a potential root cause or the problem.

example of tool application

Lack of Adherance to World Class Practices:
Quality Function Deployments (QFD), Design of
Experiments (DOE), Cycle Time Management (CTM)

Party X / Y Party	Known		Known‾
	Practiced	**Practiced‾**	
Known — **Practiced**	A	B	C
Known — **Practiced‾**	B	B	C
Known‾	C	C	D

Note: Known: WC practices of QFD, DOE, and CTM are **known** to an organizational unit.
        Practiced: WC practices of QFD, DOE, and CTM are **practiced** by an organizational unit.
        Practiced‾: WC practices of QFD, DOE, and CTM are **not practiced** by an organizational unit.
        Known‾: WC practices of QFD, DOE, and CTM are **unknown** to an organizational unit.

Appendix A: Major TQM Tools, Methods, and Processes

Definition of terms

Benchmarking/trending (BM/T) The search for industry-best practices that lead to superior performance. A method of measuring your processes against those of recognized leaders. Also a futuring process that includes trendline analyses of competitor performance data in order to determine what resources and actions are necessary to close existing gaps on an established target date. Continuous benchmarking/trending as part of the strategic planning process will eventually lead to a competitive advantage in the marketplace.

Business process reengineering (BPR) The reinventing of processes to produce significant improvements in quality, service, time, and cost. BPR disregards established practices and rules; it redesigns processes that provide extraordinary results for customers and stakeholders. According to Michael Hammer (1990), "Reengineering is the fundamental analysis and radical redesign of business processes to achieve dramatic improvements in critical performance measures."

Change factors (CF) The driving and hindering forces, causes and effects, events and pitfalls that are found in any organizational change process. The astute analysis of these factors by the change agent greatly enhances institutionalization of change.

Concurrent simultaneous engineering (C/SE) A method for integrating functional disciplines such as marketing, design engineering, manufacturing, etc. Also known as integrated product development (IPD). A systematic approach to product design that considers all elements of the product life cycle. A set of methods, techniques, and practices that: (1) causes significant consideration to be given within the design phases to

factors occurring later in the life cycle; (2) produces the production unit; (3) facilitates the reduction of the time required to translate designs into the fielded products; and (4) enhances the ability of products to satisfy users' expectations and needs.

Continuous process improvement (CPI) A systematic method to simplify and optimize any business or work-level process. Typically, an existing process is mapped to promote better understanding, performance data is collected and analyzed, and potential solutions are evaluated for possible implementation.

Customer satisfaction analysis (CSA) A customer data collection and analysis technique that summarizes qualitative and quantitative response data for the identification of improvement opportunities.

† **Cycle time management (CTM)** The ongoing analysis of work to reduce the total time required for a worker to complete one cycle or operation, including set-up, movement, delay time, and inspections. Non-value-adding time and bottlenecks are identified for elimination during this analysis.

Data collection strategy (DCS) A systematic process to identify data requirements, appropriate data collection methods to be used, and to determine what descriptive statistics are to be performed to support the problem-solving or process improvement effort.

† **Design for manufacturing assembly (DFMA)** A modern manufacturing approach that develops design criteria so that products can be designed to minimize the fabrication and assembly costs, and also improve ease of production.

Design of experiments (DOE) Planned, structured, and organized observation of two or more input/independent variables (factors) and their effect on the

---

† Indicates major manufacturing tools.

output/dependent variable(s) under study. Taguchi Methods/Analysis: techniques combining engineering and statistical methods to achieve rapid improvements in cost and quality by optimizing product design and manufacturing processes. Processes include loss function, parameter design, and signal-to-noise ratio.

Design to cost (DTC) A design engineering cost management process that requires continuous attention to meet established cost design parameters in the early product development phases.

† **Effective design review (EDR)** A systematic review of proposed design by subject matter experts with the goal of uncovering design of manufacturing deficiencies prior to implementation.

Empowerment (culture) Empowered employees have shared responsibility, risk-taking, and ownership in the decision-making and continuous improvement process throughout the organization.

Management by policy (MBP) Also known as Hoshin Planning, this planning system allows the organization's policy and focus to flow down into the organization with a prime concern for quality inter- and intradepartmental functions and applied problem-solving techniques in order to achieve a strategic organizational breakthrough.

Metrics Quantitative and qualitative metrics (measurements) in four areas of interest: resource, process, results, and customer satisfaction.

Innovative thinking A process of utilizing various tools and techniques to be more creative, generate new ideas, create opportunities, improve old processes, and move out of old thought patterns to problem solve or develop new products and services.

Integrated product development (IPD) A strategy for the management of integrating vertical and horizontal functions and processes within a TQM environment to provide an efficient and effective product/service to the customer's complete satisfaction.

ISO-9000 International Organization for Standardization is a series of international quality standards that includes documentation and conformity assessment for design, development, production, installation, and service. It is a new work ethic as well as a requirement for doing business on an international basis.

† **Just-in-time (JIT)** A concept for providing material as needed. Application of JIT management techniques reduces inventories and work in progress, lessens amount of material requiring rework, improves problem-solving, and decreases labor requirements. As a result, productivity and quality are enhanced and costs are reduced. JIT requires confidence that the supplier chain will meet commitments.

† **Poka-yoke (P-Y)** Stands for "fool proof mechanism," a concept that provides mistake proofing for supervisors and work-cell operators. It eliminates troubles associated with defects, safety, error in operation, etc., without a great amount of operator self-inspection.

Problem-solving (P-S) A process of moving from symptom to cause to action that improves process or product performance. Usually a six step, disciplined set of procedures using appropriate tools for problem identification and selection, data collection and analysis, and solution evaluation and implementation.

† **Process mapping (PM)** Identifies and documents (maps) all cross-functional processes, stakeholders, measures, flow, and type of activities. Provides input data for process improvement.

† **Process quality indicator (PQI)** PQI methods can commonly be applied to all processes. PQI methodology suggests five fundamental steps: (1) process examination; (2) data collection, recording, and analysis; (3) identification of the most important discrepancy; (4) establishment of root causes; and (5) improvement action.

Quality function deployment (QFD) A conceptual matrix (house of quality) that provides the means for cross-functional planning and communication. A method for transforming customer wants and needs (voice of the customer) into quantitative, engineering terms (voice of the company).

Self-directed work teams (SDWT) A trained team of employees who are empowered and have the accountability to manage themselves and their work.

† **Statistical process control (SPC)** Use of statistical techniques such as control charts to analyze a process or its outputs in order to maintain state of statistical control and improve the process capability.

† **Statistical quality control (SQC)** A set of powerful tools that utilizes statistical techniques to gather, analyze, and interpret information which is necessary to control or assure levels of quality required by management. SQC application monitors process output.

† **Six sigma (6σ)** Six sigma quality can be defined at two different yet closely related levels. First, at the managerial level, six sigma is a customer-driven improvement process that reflects the framework for managing quality throughout the organization. At the operational level, six sigma can be linked directly to the measurement and statistical reporting of quantitative and qualitative metrics. With six sigma, a virtual "zero defects" approach to quality, for every million opportunities to create a defect, only 3.4 defects may occur (a yield of 99.99966 percent).

Survey research (SR) A research method where data is collected from a sample of an identified population, tested, and analyzed, using profiling, descriptive, and inferential statistical techniques. Results are generalized to the population.

† **Synchronous workshop (SW)** A team approach to identify significant improvements within a relatively short time. SW will determine non-value-added tasks,

excess inventory, and waste, and will eliminate bottlenecks to reduce process cycle time, floor space, equipment, materials, and excessive movement.

† **Takt time management (TTM)** The synchronization of the manufacturing output rate to the customer demand rate in order to avoid overproduction. The takt time is calculated by determining the daily operating time and dividing it by the total daily requirement.

† **Total productive maintenance (TPM)** A process that smooths the flow of material through the manufacturing process by continually improving production machine efficiency and reliability. This is achieved by improving equipment effectiveness, developing a planned maintenance program, training machine operators and allowing them to perform autonomously, and instituting a program to prevent problems before they occur.

Total quality management (TQM) Consists of continuous process improvement activities involving everyone in an organization—managers and workers—in a totally integrated effort toward improving performance at every level. This improved performance is directed toward satisfying such cross-functional goals as quality, cost-schedule, mission, need, and suitability. TQM integrates fundamental management techniques, existing improvement efforts, and technical tools under a disciplined approach focused on continuous process improvement. The activities are ultimately focused on increased customer/user satisfaction.

† **Value-added manufacturing (VAM)** The implementation of three basic elements encompassing just-in-time manufacturing, total commitment to quality, and employee involvement.

Value engineering (VE) A systematic analysis of all functions deemed necessary to remove costs in the overall design process of products and services.

† **Variability reduction process (VRP)** The desired result of the application of statistical and analytical methods

to achieve continuous measurable improvement of process, products, and service. These methods include, but are not limited to, quality function deployment (OFD), design of experiments (DOE), statistical quality/process control (SQC/SPC), and the seven management tools.

† **Vendor quality assurance (VQA)** A formal program that emphasizes the need for increasing the excellence of items and services we purchase. Could include supplier training and certification.

† **Work redesign and simplification (WRS)** A process of changing the current work activities in order to:

(1) enhance job satisfaction by enriching the work setting and tasks performed: (2) improve quality and productivity by increasing effectiveness and efficiency; and (3) provide more meaningful work that benefits both the individual and the organization. Five primary principles involved are to change the method, sequence, equipment, material, or design of the present or planned work.

† **World class manufacturing (WCM)** Reflects an overriding goal of continual and rapid improvement in quality, cost, lead time, and customer service.

Appendix B: Cross-Reference Index

This cross-reference alphabetically lists the 222 tools (as well as their aka's) as they appear in *Tool Navigator™*. Tool cross-references are in parentheses. Numerical designations, for example number **122**, indicate the tool number. Letter designations, for example (IG), indicate tool classification (Idea Generating). The tools in bold type include the tools in this book, as well as other manufacturing and supporting tools that may be typically used by manufacturing.

Tool Number in *40 Top Tools for Manufacturers*	Tool Number in *Tool Navigator™*		Tool Name
	1	(IG)	5W2H method
	2	(IG)	6-3-5 method
			Abstraction process (*see* tool 102)
			Accountability grid (*see* tool 166)
			Action and consequences diagram (*see* tool 86)
	3	(AT)	Action and effect diagram (AED)
	4	(PP)	Action plan
12	5	(AT)	**Activity analysis**
	6	(DC)	**Activity cost matrix**
1	7	(PP)	**Activity network diagram**
			Affinity analysis (*see* tool 8)
	8	(IG)	Affinity diagram
	9	(IG)	Analogy and metaphor
16	10	(AT)	**Analysis of variance**
			ANOVA, *F*-test (*see* tool 10)
			Area graph (*see* tool 133)
			Arrow analysis (*see* tool 7)
			Assembly flow (*see* tool 221)
	11	(IG)	Attribute listing
	12	(PP)	Audience analysis
	13	(DM)	Balance sheet
			Bar graph (*see* tool 14)
	14	(AT)	**Bar chart**
	15	(CI)	Barriers-and-aids analysis
	16	(PP)	**Basili data collection method**
	17	(DC)	Benchmarking
			Benefits and barriers exercise (*see* tool 15)
	18	(AT)	Block diagram
			Box and whisker plot (*see* tool 19)
	19	(AT)	**Box plot**
			Brain webs (*see* tool 110)
	20	(IG)	Brainstorming
			Brainwriting (*see* tool 21)
	21	(IG)	**Brainwriting pool**
	22	(AT)	Breakdown tree
			Bulletin board fishbone (*see* tool 26)
	23	(TB)	Buzz group
			Buzzing (*see* tool 23)
			Capability assessment chart (*see* tool 126)
			Capability indices (*see* tool 147)
			Case analysis method (*see* tool 24)
	24	(DC)	Case study
			Cause analysis (*see* tool 143)
	25	(AT)	**Cause and effect diagram (CED)**
28	26	(AT)	**Cause and effect diagram adding cards (CEDAC)**
			Checkerboard diagram (*see* tool 27)
	27	(IG)	Checkerboard method
2	28	(IG)	**Checklist**
29	29	(DC)	**Checksheet**
			Chi-square analysis (*see* tool 92)
			Circle chart (*see* tool 133)
	30	(IG)	Circle of opportunity
	31	(DC)	**Circle response**

Tool Number in *40 Top Tools for Manufacturers*	Tool Number in *Tool Navigator*™		Tool Name
	32	(TB)	Circles of influence
	33	(DC)	Circles of knowledge
	34	(IG)	Circumrelation
	35	(ES)	Cluster analysis
			Clustering (*see* tool 35)
			Color code audit (*see* tool 103)
			Color dots rating (*see* tool 187)
			Comparative assessment matrix (*see* tool 36)
			Comparative benchmarking (*see* tool 17)
			Comparison grid (*see* tool 128)
	36	(ES)	Comparison matrix
	37	(AT)	Competency gap assessment
			Concentration diagram (*see* tool 60)
	38	(DC)	Conjoint analysis
	39	(DM)	Consensus decision making
			Consensus generator (*see* tool 39)
17	40	(AT)	**Control Chart - *c* (attribute)**
18	41	(AT)	**Control Chart - *p* (attribute)**
19	42	(AT)	**Control Chart - $\bar{X}$-R (variable)**
	43	(DM)	Correlation analysis
20	44	(AT)	**Cost of quality**
			Cost of quality analysis (*see* tool 44)
	45	(ES)	Cost-benefit analysis
	46	(PP)	**Countermeasures matrix**
	47	(IG)	Crawford slip method
			Crawford slip writing (*see* tool 47)
			Creative evaluation (*see* tool 48)
	48	(ES)	Creativity assessment
	49	(ES)	**Criteria filtering**
			Criteria ranking (*see* tool 159)
			Criteria rating (*see* tool 160)
			Criteria rating form (*see* tool 135)
	50	(IG)	Critical dialogue
	51	(DC)	Critical incident
			Critical path method (CPM) (*see* **tool 152**)
			Cross-functional matrix (*see* tool 85)
			Cross-functional process map (*see* tool 150)
	52	(AT)	Customer acquisition-defection matrix
	53	(DC)	Customer needs table
	54	(AT)	Customer satisfaction analysis (CSA)
			Customer satisfier matrix (*see* tool 54)
			Customer window grid (*see* tool 178)
	55	(DC)	**Customer-first-questions (CFQ)**
13	56	(AT)	**Cycle time flowchart**
			Daily operations log (*see* tool 71)
			Data collection plan (*see* tool 57)
	57	(DC)	**Data collection strategy**
			Data entry form (*see* tool 164)
			Data matrix (*see* tool 164)
			Decision flow analysis (*see* tool 59)
			Decision model (*see* tool 58)
	58	(DM)	Decision process flowchart
	59	(DM)	Decision tree diagram
			Defect location checksheet (*see* tool 60)
30	60	(DC)	**Defect map**
	61	(ES)	Delphi method
	62	(CI)	**Deming PDSA cycle**
			Deming wheel (*see* tool 62)
	63	(AT)	Demographic analysis
31	64	(ES)	**Dendrogram**
	65	(PP)	Deployment chart (down-across)
	66	(PP)	**Descriptive statistics**
			Dialogue (*see* tool 50)
	67	(PP)	Different point of view
			Dimension map (*see* tool 157)

Tool Number in *40 Top Tools for Manufacturers*	Tool Number in *Tool Navigator™*		Tool Name
	68	(PP)	Dimensions cube
			Direct association (*see* tool 81)
			Distribution ratio (*see* tool 133)
	69	(AT)	Dot diagram
			Dot plot (*see* tool 69)
	70	(IG)	Double reversal
21	**71**	**(CI)**	**Events log**
3	**72**	**(CI)**	**Facility layout diagram**
	73	**(ES)**	**Factor analysis**
32	**74**	**(CI)**	**Failure mode effect analysis (FMEA)**
33	**75**	**(AT)**	**Fault tree analysis (FTA)**
			Fishbone diagram (*see* tool 25)
	76	(TB)	Fishbowls
			Five w's and two h's (*see* tool 1)
	77	(AT)	Five whys
			Flow methods diagram (*see* tool 194)
	78	(DC)	Focus group
	79	(CI)	Fog index
	80	(CI)	Force field analysis (FFA)
	81	(IG)	Forced association
	82	(DM)	Forced choice
			Forced comparison (*see* tool 82)
			Forced relationship method (*see* tool 34)
			Free association (*see* tool 30)
			Free-form brainstorming (*see* tool 20)
22	**83**	**(AT)**	**Frequency distribution (FD)**
			Frequency table (*see* tool 83)
	84	(IG)	Fresh eye
			Function block diagram (*see* tool 18)
	85	(AT)	Functional map
	86	(AT)	Futures wheel
			Futuring (*see* tool 86)
			Gallery method (*see* tool 21)
	87	(PP)	Gantt chart
			Gantt planning (*see* tool 87)
	88	(CI)	Gap analysis

Tool Number in *40 Top Tools for Manufacturers*	Tool Number in *Tool Navigator™*		Tool Name
			Gaussian curve (*see* tool 119)
			Goal planning (*see* tool 16)
4	89	(PP)	**Gozinto chart**
			Gunning fog index (*see* tool 79)
	90	(AT)	**Histogram**
			Histogram analysis (*see* tool 90)
	91	(ES)	House of quality
			Hypothesis testing (ANOVA) (*see* tool 10)
	92	(DM)	Hypothesis testing (chi-square)
			Hypothesis testing (correlation) (*see* tool 43)
	93	(ES)	Idea advocate
	94	(IG)	Idea borrowing
			Idea box (*see* tool 112)
	95	(IG)	Idea grid
			Imagery (*see* tool 108)
			Impact-effort analysis (*see* tool 49)
	96	(DM)	Importance weighting
			Incident analysis (*see* tool 51)
	97	(PP)	Influence diagram
	98	(DC)	Information needs analysis
			Input-output analysis (*see* tool 195)
			Instant priorities (*see* tool 120)
			Interaction diagram (*see* tool 182)
			Interaction-relations diagram (*see* tool 99)
	99	(AT)	Interrelationship digraph (I.D.)
	100	(DC)	Interview technique
			Interviewing (*see* tool 100)
			Ishikawa diagram (*see* tool 25)
			Job analysis (*see* tool 196)
			Jury of experts (*see* tool 61)
			K-J method (*see* tool 8)
			Line graph (*see* tool 101)
23	**101**	**(AT)**	**Line chart**
			Link analysis (*see* tool 221)

Tool Number in *40 Top Tools for Manufacturers*	Tool Number in *Tool Navigator*™		Tool Name
	102	(ES)	**Linking diagram**
			List reduction (*see* **tool 49**)
			Logic diagram (*see* tool 58)
	103	(CI)	Major program status
	104	(AT)	Markov analysis
			Matrix chart (*see* tool 106)
	105	(ES)	Matrix data analysis
34	106	(PP)	**Matrix diagram**
	107	(ES)	Measurement matrix
			Meeting process check (*see* tool 197)
	108	(IG)	Mental imaging
			Metaphorical thinking (*see* tool 9)
	109	(PP)	Milestones chart
	110	(IG)	Mind flow
14	111	(IG)	**Monthly assessment schedule**
	112	(IG)	Morphological analysis
			Morphological forced connections (*see* tool 112)
			Multiple line graph (*see* tool 190)
	113	(DC)	Multiple rating matrix
			Multiple rating profile (*see* tool 113)
			Multi-var chart *see* tool 114)
24	114	(AT)	**Multivariable chart**
			Multi-vote technique (*see* tool 115)
	115	(DM)	Multivoting
	116	(AT)	Needs analysis
			Needs assessment (*see* tool 116)
			Network diagram (*see* tool 99)
			Node diagram (*see* tool 7)
			Nominal group process (*see* tool 117)
	117	(IG)	Nominal group technique (NGT)
			Nominal grouping (*see* tool 117)
	118	(ES)	**Nominal prioritization**
	119	(AT)	**Normal probability distribution**
	120	(ES)	Numerical prioritization
	121	(CI)	**Objectives matrix (OMAX)**

Tool Number in *40 Top Tools for Manufacturers*	Tool Number in *Tool Navigator*™		Tool Name
	122	(DC)	Observation
	123	(ES)	Opportunity analysis
	124	(PP)	Organization chart
	125	(TB)	Organization mapping
	126	(CI)	Organization readiness chart
			Organizational mirror (*see* tool 198)
			Osborne brainstorming (*see* tool 20)
			Outsider's view (*see* tool 67)
	127	(TB)	Pair matching overlay
	128	(ES)	Paired comparison
			Pairwise ranking (*see* tool 120)
	129	(AT)	Panel debate
			Pareto analysis (*see* tool 130)
35	130	(AT)	**Pareto chart**
			Pareto principle (*see* tool 130)
			Partner link (*see* tool 127)
			Percent change bar graph (*see* tool 209)
			Performance gap analysis (*see* tool 37)
			Performance index (*see* tool 121)
			Perspective wheel (*see* tool 169)
	131	(IG)	Phillips 66
			Phillips 66 buzz session (*see* tool 131)
			Pictogram (*see* tool 132)
	132	(PP)	**Pictograph**
	133	(AT)	Pie chart
	134	(IG)	Pin cards technique
			Plan-do-check-act strategy (*see* tool 180)
			Planning schedule (*see* tool 109)
	135	(DM)	Point-scoring evaluation
	136	(AT)	**Polygon**
			Polygon analysis (*see* tool 136)
	137	(AT)	**Polygon overlay**
			Polygon trend comparison (*see* tool 137)

Tool Number in *40 Top Tools for Manufacturers*	Tool Number in *Tool Navigator*™		Tool Name
			Positive-negative chart (*see* tool 209)
36	138	(CI)	**Potential problem analysis (PPA)**
	139	(PP)	Presentation
			Presentation review (*see* tool 139)
	140	(ES)	**Prioritization matrix—analytical**
	141	(ES)	**Prioritization matrix—combination**
	142	(ES)	**Prioritization matrix—consensus**
	143	(AT)	**Problem analysis**
			Problem definition (*see* tool 145)
	144	(ES)	**Problem selection matrix**
			Problem solution planning (*see* tool 46)
37	145	(PP)	**Problem specification**
5	146	(AT)	**Process analysis**
25	147	(AT)	**Process capability ratios**
			Process cycle time analysis (*see* tool 56)
	148	(PP)	Process decision program chart (PDPC)
			Process flow analysis (*see* tool 149)
6	149	(AT)	**Process flowchart**
7	150	(CI)	**Process mapping**
8	151	(CI)	**Process selection matrix**
	152	(PP)	**Program evaluation and review technique (PERT)**
	153	(PP)	Project planning log
	154	(ES)	Project prioritization matrix
			Pros and cons (*see* tool 13)
	155	(CI)	**Quality chart**
	156	(DC)	Questionnaires
	157	(AT)	Radar chart
	158	(DC)	Random numbers generator
			Random numbers table (*see* tool 158)
	159	(ES)	Ranking matrix
	160	(ES)	Rating matrix
	161	(TB)	Relationship map
			Requirements matrix (*see* tool 163)
			Requirements QFD matrix (*see* tool 91)

Tool Number in *40 Top Tools for Manufacturers*	Tool Number in *Tool Navigator*™		Tool Name
162		(PP)	**Resource histogram**
			Resource loading (*see* tool 162)
163		(PP)	**Resource requirements matrix**
164		(DC)	Response data encoding form
165		(DC)	Response matrix analysis
166		(CI)	Responsibility matrix
			Results reporting (*see* tool 107)
167		(ES)	Reverse brainstorming
			Reverse fishbone (*see* tool 3)
			Reversed thinking (*see* tool 70)
			Reviewed dendrogram (*see* tool 64)
			Risk management diagram (*see* tool 168)
168		(AT)	Risk space analysis
			Root cause analysis (*see* tool 77)
			Rotating chairs (*see* tool 169)
169		(CI)	Rotating roles
170		(IG)	Round robin brainstorming
			Recorded round robin technique (*see* tool 2)
171		(AT)	**Run chart**
172		(AT)	Run-it-by
			Sample analysis (*see* tool 22)
			Sampling (random/systematic/stratified/cluster) (*see* tool 173)
			Sampling chart (*see* tool 158)
173		(DC)	**Sampling methods**
			Sarape chart (*see* tool 191)
38	174	(AT)	**SCAMPER**
			SCAMPER questions (*see* tool 174)
			Scatter analysis (*see* tool 175)
175		(AT)	**Scatter diagram**
			Scatterplot (*see* tool 175)
			Scenario construction (*see* tool 176)

135

Tool Number in *40 Top Tools for Manufacturers*	Tool Number in *Tool Navigator*™		Tool Name
	176	(CI)	Scenario writing
			Second opinion (*see* tool 172)
			Selection grid (*see* tool 177)
	177	**(ES)**	**Selection matrix**
	178	(ES)	Selection window
	179	(IG)	Semantic intuition
	180	**(CI)**	**Shewhart PDCA cycle**
			Situation analysis (*see* tool 193)
	181	(AT)	Snake chart
	182	(TB)	Sociogram
			Sociometric diagram (*see* tool 182)
			Solution impact diagram (*see* tool 3)
39	**183**	**(ES)**	**Solution matrix**
			Solutions selection matrix (*see* tool 183)
			Space plot (*see* tool 168)
			Spider chart (*see* tool 157)
	184	**(AT)**	**Standard deviation**
	185	(DC)	Starbursting
			Stem-and-leaf diagram (*see* tool 186)
	186	(AT)	Stem-and-leaf display
	187	(ES)	Sticking dots
	188	(IG)	Stimulus analysis
			Storyboard (*see* tool 189)
	189	(PP)	Storyboarding
	190	**(AT)**	**Stratification**
26	**191**	**(AT)**	**Stratum chart**
			Surface chart (*see* tool 191)
			Survey analysis (*see* tool 192)
			Survey profiling (*see* tool 165)
	192	(DC)	Surveying
	193	(PP)	SWOT analysis
	194	**(AT)**	**Symbolic flowchart**
			Systematic diagram (*see* tool 204)
	195	**(AT)**	**Systems analysis diagram**
			Tally sheet (*see* tool 29)
	196	**(AT)**	**Task analysis**
	197	(TB)	Team meeting evaluation
	198	(TB)	Team mirror
	199	(TB)	Team process assessment
			Team rating (*see* tool 187)
			Teardown method (*see* tool 167)
	200	(AT)	Thematic content analysis
			Time plot (*see* tool 202)
			Time series analysis (*see* tool 205)
15	**201**	**(CI)**	**Time study sheet**
	202	(AT)	Timeline chart
9	**203**	**(PP)**	**Top-down flowchart**
			Traditional organization chart (*see* tool 124)
			Tree analysis (*see* tool 204)
	204	(PP)	Tree diagram
	205	**(AT)**	**Trend analysis**
	206	(ES)	Triple ranking
	207	(AT)	Truth table
			Two-dimensional scatter diagram (*see* tool 208)
	208	(DC)	Two-dimensional survey grid
	209	(AT)	Two-directional bar chart
	210	(AT)	Value analysis
	211	**(AT)**	**Value/non-value-added cycle time chart**
27	**212**	**(AT)**	**Variance analysis**
			Variance matrix (*see* tool 212)
	213	(ES)	Venn diagram
			Visualization (*see* tool 108)
			Voice of the customer (*see* tool 53)
			Weighted averages matrix (*see* tool 96)
	214	(DM)	Weighted voting
	215	(AT)	What-if analysis
	216	(PP)	Why/how charting
	217	(IG)	Wildest idea technique
			Wildest idea thinking (*see* tool 217)
40	**218**	**(AT)**	**Window analysis**
	219	(AT)	Wishful thinking
			Work breakdown diagram (*see* tool 220)
10	**220**	**(PP)**	**Work breakdown structure (WBS)**
11	**221**	**(CI)**	**Work flow analysis (WFA)**
	222	**(AT)**	**Yield chart**

Abbreviated Bibliography

Aft, Lawrence S. *Productivity, Measurement, and Improvement*. Milwaukee, WI: Quality Press, 1983.

Alsup, Fred and Ricky M. Watson. *Practical Statistical Process Control*. New York: Van Nostrand Reinhold, 1993.

Born, Gary. *Process Management to Quality Improvement*. New York: John Wiley & Sons, Inc., 1994.

Burr, Irving W. *Statistical Quality Control Methods*. Milwaukee, WI: ASQC Quality Press, 1979.

Carruba, Eugene R. and Ronald D. Gordon. *Product Assurance Principles: Integrating Design Assurance and Quality Assurance*. Milwaukee, WI: Quality Press, 1988.

Christopher, William F. *Productivity Measurement Handbook* (2nd ed.). Portland, OR: Productivity Press, 1985.

Davenport, Thomas H. *Process Variation*. Boston, MA: Harvard Business School Press, 1993.

Electronic Systems Division. *The ESD Process Improvement Guide*. Hanscom AFB, MA: ESD, MITRE, 1991.

Feigenbaum, Armand V. *Total Quality Control* (3rd ed.). New York: McGraw-Hill, Inc., 1983.

Ford Motor Company. *Continuing Process Control and Process Capability Improvement*. Dearborn, MI: Statistical Methods Office, 1984.

Grant, Eugene L. and Richard S. Leavenworth. *Statistical Quality Control* (6th ed.). New York: McGraw-Hill, Inc., 1988.

Hayes, Glenn E. *Quality Assurance: Management and Technology* (rev. ed.). Capistrano Beach, CA: Charger Productions, 1983.

Ishikawa, Kaoru. *Guide to Quality Control*. Tokyo: Asian Productivity Organization, 1986.

Ishikawa, Kaoru. *Introduction to Quality Control*. Tokyo: 3A Corporation, 1989.

Jones, Morgan D. *The Thinker's Toolkit*. New York: Random House, Inc., 1995.

Kinlaw, Dennis C. *Continuous Improvement and Measurement for Total Quality*. Homewood, IL: Pfeiffer & Company, 1992.

Levin, Richard I. and David S. Rubin. *Statistics for Management* (5th ed.). Englewood Cliffs, NJ: Prentice-Hall, Inc., 1991.

Levine, Marvin. *Effective Problem Solving* (2nd ed.). Englewoods Cliffs, NJ: Prentice-Hall, Inc., 1994.

Mears, Peter. *Quality Improvement Tools & Techniques*. New York: McGraw-Hill, Inc., 1995.

Messina, William S. *Statistical Quality Control for Manufacturing Managers*. New York: John Wiley & Sons, Inc., 1987.

Michalski, Walter J. *Continuous Measurable Improvement Poster (99 Tools)*. Fullerton, CA: Hughes Aircraft Company, 1992.

Michalski, Walter J. *Tools and Notes*. Unpublished collection of approximately 500 tools and techniques for quality and process improvement, collection period 1965–1997.

Michalski, Walter J. *Tools for Teams: Hands-On Problem Solving and Process Improvement Manual*. Huntington Beach, CA: Alpha Research Group, 1994.

Robson, George D. *Continuous Process Improvement*. New York: The Free Press, 1991.

Rubinstein, Moshe F. *Tools for Thinking and Problem Solving*. Englewood Cliffs, NJ: Prentice-Hall, Inc., 1986.

Ryan, Thomas P. *Statistical Methods for Quality Improvement*. New York: John Wiley & Sons, 1989.

Scholtes, Peter R. *The Team Handbook*. Madison, WI: Joiner Associates, Inc., 1988.

Shores, Richard A. *A TQM Approach to Achieving Manufacturing Excellence*. Milwaukee, WI: ASQC Quality Press, 1990.

Swanson, Roger C. *The Quality Improvement Handbook*. Delray Beach, FL: St. Lucie Press, 1995.

Tague, Nancy R. *The Quality Toolbox*. Milwaukee, WI: ASQC Quality Press, 1995.

Timko, John J. *Statistics by Example* (2nd ed.). Orange, CA: Statistics for Management, 1993.

Van Gundy, Arthur B. *Techniques of Structured Problem-Solving*. New York: Van Nostrand Reinhold Company, 1981.

Wheeler, Donald J. *Understanding Variation*. Knoxville, TN: SPC Press, Inc., 1993

Xerox Corporation. *Problem-Solving Process User's Manual*. Rochester, NY: Business Products Systems Group, 1987.

About the Author

Dr. Walter J. Michalski is president of Alpha Research Group, Huntington Beach, California, a TQM consulting firm that assists organizations in their quality and change initiatives. His work experience reflects 28 years in quality assurance, test/process engineering, and process improvement training. He has designed many problem-solving workshops and facilitated teams on quality issues, process reengineering, continuous improvement, measurements, organizational change, and critical-thinking techniques. He holds an Ed.D. in Institutional Management from Pepperdine University, GSEP, Los Angeles, CA.; his doctoral dissertation examined the effectiveness of nontraditional college degree programs. As an adjunct professor, he continues to teach at graduate/undergraduate levels in subjects such as research methods, statistics, TQM/TQS, management and organizational behavior, and he serves as project advisor for students' research projects, practicums, and theses. He can be reached at Alpha Research Group, 8481 Ivy Circle, Huntington Beach, CA 92646, (714) 968-0452, e-mail: 103217.1070@compuserve.com.